"Beresford-Kroeger is a lyrical storyteller. . . . [She] declares that forests hold the solution to climate change as well as countless cures for a multitude of ailments including cancer." —*ON Nature*

"In her Irish-Gaelic storytelling voice, Beresford-Kroeger hooks her audience by writing in a way they can understand. She leads them into this forested world by connecting two seemingly unrelated topics, then laces these two ideas together, exposing her audience to deeper needs: the cry of planet earth and humanity's responsibility to that cry." —*New World Review*

"Through this compelling series of forty essays, Beresford-Kroeger builds a continuum that is at once simplistic and profound, a style, I dare say, not unlike that found in the texts of mankind's various religions. . . . This book is a must-read, especially for those who remain skeptical at the need to beware the evil twins of global warming and climate change." —*Winnipeg Free Press*

"Mixes mythology with science and poetry with factual detail on virtually every page . . . the science supports the poetry in ways that may well make both appear invincible. . . . The book may well resonate with a wide variety of readers for many years to come." —*The Taipei Times*

"The book's forty short chapters delve into the intrinsic value of trees, from their medicinal and carbon-sucking properties to their spiritual connection with cultures throughout history. . . . Like the slow-moving quietude of an old-growth forest, this book is meant to gradually sink into your soul . . . [laying] down subtle and piercing roots that eventually strengthen and hold firm." —*Canadian Geographic Magazine*

"I have never read anything quite like *The Global Forest*. It reads like a spinning continuum of a mind-altering walk through the most magnificent forests in the world, looking back into the deepest possible crannies of a not-so-distant human history and consciousness while also seeing ahead into the near future of science and consequence, with the hair on the back of one's neck rising, yet looking forward also with candle glimmers of hope and yearning that cannot be vanquished." —Rick Bass

PENGUIN BOOKS

THE GLOBAL FOREST

Diana Beresford-Kroeger is a botanist and medical biochemist who is an expert on the medicinal, environmental, and nutritional properties of trees. Her previous books include *Arboretum Borealis: A Lifeline of the Planet*; *A Garden for Life: The Natural Approach to Designing, Planting, and Maintaining a North Temperate Garden*; and *Arboretum America: A Philosophy of the Forest*. She lives in Ontario, Canada, surrounded by her sprawling research gardens filled with rare and endangered species.

The Global Forest

Forty Ways Trees Can Save Us

DIANA BERESFORD-KROEGER

PENGUIN BOOKS

PENGUIN BOOKS

Published by the Penguin Group

Penguin Group (USA) Inc., 375 Hudson Street, New York, New York 10014, U.S.A. • Penguin Group (Canada), 90 Eglinton Avenue East, Suite 700, Toronto, Ontario, Canada M4P 2Y3 (a division of Pearson Penguin Canada Inc.) • Penguin Books Ltd, 80 Strand, London WC2R 0RL, England • Penguin Ireland, 25 St. Stephen's Green, Dublin 2, Ireland (a division of Penguin Books Ltd) • Penguin Books Australia Ltd, 250 Camberwell Road, Camberwell, Victoria 3124, Australia (a division of Pearson Australia Group Pty Ltd) • Penguin Books India Pvt Ltd, 11 Community Centre, Panchsheel Park, New Delhi – 110 017, India • Penguin Group (NZ), 67 Apollo Drive, Rosedale, Auckland 0632, New Zealand (a division of Pearson New Zealand Ltd) • Penguin Books (South Africa) (Pty) Ltd, 24 Sturdee Avenue, Rosebank, Johannesburg 2196, South Africa

Penguin Books Ltd, Registered Offices: 80 Strand, London WC2R 0RL, England

First published in the United States of America by Viking Penguin,
a member of Penguin Group (USA) Inc. 2010
Published in Penguin Books 2011

THE LIBRARY OF CONGRESS HAS CATALOGED THE HARDCOVER EDITION AS FOLLOWS:
Beresford-Kroeger, Diana, 1944–
    The global forest / Diana Beresford-Kroeger
        p.   cm.
    Includes bibliographical references.
    ISBN 978-0-670-02174-1 (hc.)
    ISBN 978-0-14-312016-2 (pbk.)
    1. Trees—Folklore.   2. Trees—Symbolic aspects.   3. Forest ecology.   I. Title.
    GR785.B47 2010
    398.24'2—dc22   2009046313
Printed in the United States of America
Designed by Carla Bolte • Set in Berkeley Oldstyle

TO ERIKA

A palm so small
it fits in mine,
Mother and Child
of the universe.
You are of me
a song of love,
you sit beside
all that is beautiful
in you to be.

CONTENTS

❧ THE GLOBAL FOREST ❧

INTRODUCTION

The landscape of my youth was an Irish one. The fields were filled with the brilliant chrome yellow of furze. In between these bushes the grasses grew that softly sang the green of a perennial life. There never was one word spoken to remove those bushes and make way for more pasture. The furze was part of nature, part of the whole that worked together to sustain both man and beast.

I was small enough to squeak through the holes in the hedges to follow the horses and the one brave donkey that endured my affections. The cows were there, too, with a handful of sheep. They were all out in the morning pasture that they loved. This field was large and steep. It rolled down to a stream that ran out of the foot of the field amid some outcrops of black rocks. This field gathered up all of the morning sun and the grasses held all of the morning dew.

I knew what the horses and the donkey would do. They would approach the furze bushes. They would dip their noses down into the bright yellow blooms and smell along the cropped line of the bush until they got to the spot they wanted. Then they would pull back their upper and lower lips in a grimace, exposing a double row of long browning teeth. They would open these teeth and continue the contour of the browse line with a sharp neat snip. This was followed by the comfortable ruminations of a green breakfast and all of the soft sounds of swallowing and gentle mobile breath. Tails would flick at nothing, always in a cascade of unison.

I was in forbidden territory. I was five. The horses and especially the donkey were dangerous. I was told time and time again
to stay away from their feet, especially the hind legs. But I was
not interested in those legs. I was interested in the teeth and most
especially in the bushes they were so carefully eating. I got my
own reward from all that observation and it came in the form of a
field mushroom, *Agaricus campestris*, a delicious product of dung.
The bush the creatures so delightfully clawed was the furze or
common gorse, *Ulex europaeus*, the larger of the two *Ulex* species
common to Ireland's hills and valleys.

This one of the two species on the farm firmly hugged the hedgerows. The other smaller, squatter species was down near the stream
in the company of the black rocks. This dwarf version, *U. gallii*, was
more prostrate and easier for me to examine. My overwhelming
interest in these shrubs was in their yellow pealike blooms which
could be opened by a slight sideways squeeze to reveal a hidden
interior. And both shrubs had unmerciful thorns of unbelievable
sharpness. I could not bear for my small hand to rest on them, but
the horses and one donkey quite readily managed to eat with relish.
The cows did not share these same fodder feelings.

When I was still younger, somebody in the household had
brought me along while they were hanging out the week's washing. All the clothes, including my dresses, were stretched out on
these furze bushes. The weight of the damp clothes pressed down
on the thorns which pinned the clothes in place. As the clothes
dried in the sweet sea breezes off the Atlantic, my dresses were
filled with pinholes of aeration. The cotton fabric took on the look
of the furze bushes, which delighted me no end.

As the autumn sun parched the mountains, they and the distant
fields rolled into a purple bed. The tiny bells of heath, the flowers

of heather, rang the way for the scorching of brackens. The dung from the horses and donkey melted, crumble by crumble, into the green grass. And from this mixture a mushroom arose. It was huge, with wide breathing gills like a fish. When I looked down the field from my perch, there were mushrooms everywhere, brown and white stalwarts with ruggedly strong stipes and pink-fleshed buttons just blinking at the first morning light.

I gathered up the front of my dress and picked into it. I arrived back at the farm with my first feast. In the kitchen, a laugh followed by several chuckles greeted me. The mushrooms were shaken free of their dirt. A nub of homemade butter was put into each. They were placed on the bastible baking dish over an open peat fire.

Strings of knowledge connect this memory in my mind. The furze, the *Ulex* species, are members of the *Leguminosae* or pea family. This family is an avid nitrogen fixer. The horses need this nitrogen in their diet. The dung was rich in it. This, in turn, induced the field mushroom to fruit. And, then . . . I had my feast.

This was as impressive for a five-year-old as it still is to me today. The forests and our lives are connected and interconnected in ways that we can only wonder about. Sometimes science walks in and answers. Other times the answers are so simple that they are obvious in themselves. It is out of this child, this field with its golden furze, that this book was born.

But this book has experienced the hand and mind of another teacher, one that is long gone, one who was a drifter in the Irish landscape. He landed like a butterfly filled with knowledge of the ancient sort. He always came in the night, through the scented darkness of the *bótharín*, the little roadway, into the farmhouse. That man owned a piece of tradition in the turf-warmed kitchen in common with all of his kind. That property was a settee, a *leaba shuíocháin*, a hard, deal bench that was his bed and his seat by the community of the fire.

This drifter was the *Seanchaí*, or traditional storyteller. He was the keeper of legends and oral traditions of his Irish brethren. These were passed down within his family lines to share with all who would hear. He was the living memory bank of his race . . . "he was the one!" The *Seanchaí* was the most important visitor to the farmhouse. All else came after him in the pecking order. His voice held the mysteries of life itself and his riddles encased them in that ancient throne of Gaelic wisdom.

When he was fed and settled by the turf fire, the hills emptied to his heels. The local farmers came smelling of sweet hay and freshening cows with rod and perch in their brains. The mountain people came through the half-door with a windy billow of an Irish poem. They all came. They always stayed because the night, that night, would be so sweet.

The *Seanchaí* began like a wet dog, rounding his backside in the three-time circle of the wolf. He threw his idea as a refrain into the flames for it to float around, to be chewed upon, to be thought about and finally to be digested. The idea was always short, some-times in Gaelic, sometimes not. The words were carefully fed out, as the backside settled into its stride, forming the short refrain. This piece was passed along from person to person in the wonder of itself like an echo of the past into its own domain. And then the story began.

And so, each one of my stories is presented to you first as a refrain. This is for thought. Ideas are the food of the mind. Thoughts and ideas beget curiosity. Then my story begins. There are forty of them. Each is in essay form. Combined, they are called *The Global Forest*. Each leaf of every tree makes up the global forest.

This forest is the environment that drives and fulfills the dream of each leaf in a vast rhythmic cycle called life. Nothing is outside. We are all of it in a unity that transcends the whole. Maybe, just maybe, this resonates of God. If that is so, then we are all His children, every earthworm, every virus, mammal, fish and whale, every fern, every tree, every man, woman, and child. One equal to another. Again and again.

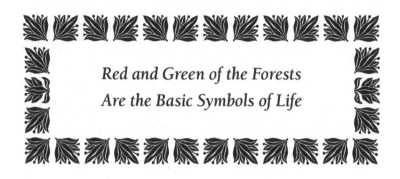

*Red and Green of the Forests
Are the Basic Symbols of Life*

BLOODLINES

Red and green have been mystical colors since ancient times. These colors were symbols for the human race long before written language. Red and green were the colors of the warrior Celts who stamped across Europe into battle naked, while at home the cult of their Druidic priests plied their sacred image in prayer. In the end, colors are like names; they are carried in the mind. Their visual simplicity allows them to be retained deep in memory. The colors red and green ride the tides of civilization with instant recall. They net in their meaning and symbolism intact, to be fed as advertising bait into our modern techno times.

There is in the global garden a very well-known tree. It is holly, *Ilex aquifolium*, an evergreen of Europe, Africa, and China. It has a deciduous cousin in North America, *I. verticillata*, winterberry, as well as the evergreen holly, *I. opaca*. They are all important medicinal trees, being used in the management of elevated fevers. The Druids had long adopted the holly for their pagan festival of light and darkness, which comes down to us today as Christmas.

During the season of Christmas, holly is now transported all over the globe. Since its old name is "holy plant," it fits in with the birth of the Christ child. This species, with its forest green leaves and tight cymes of bloodred berries, is used to decorate homes for the Christmas holidays. Even the Christmas pudding is somewhat surprised to find itself ablaze with its topknot of holly, all an echo of those pagan times.

Holly is that mystical plant of green and red. In times past the deep color represented the green of the ancient virgin forests and all of the secret powers that they held. These were considered to be holy places and for many still are. Holly achieves its forest green color by a trick of optics. The upper layer of the leaves has a waxy film that amplifies the color and gives it an optical depth. For the Druids the berry color is exactly that of fresh blood, that particular scarlet of sacrifice, human of course.

In addition, to the Celts, red and green represented the dichotomy of our lives. It was true then and it is true now. The forest green represents the plants that serve us and give us sustenance for life. And red is the deep limbic knowledge of self, the circulation and the blood that flows through us. Both systems of man and forest depend on each other.

The symbolism and perhaps even the mysticism of red and green originate at a molecular level. Blood is a red pigment that functions like an oil. It is primarily made up of flexible, mobile hemoglobin molecules contained in doughnut-shaped sacs, the red blood cells. It has genius in its design and this design is shared in a remarkable way by the green oily pigment of plants, chloroplast. The chloroplasts are also sacs containing the flexible, mobile chlorophyll molecules. These two sister molecules, hemoglobin and chlorophyll, the red and the green, conduct the pattern of our lives. Without them we would not survive as a species or as a planet.

But there is more to the molecular story. Both hemoglobin and chlorophyll are molecular machines. They work in a similar manner almost as if they were related to each other, which they are in a wider global sense. Their family kinship is built on the design they have in common, four aromatic rings that contain nitrogen. Sitting in the center position of these rings, like a solitaire diamond, is an atom of metal. In attendance to this metal, holding it in place like a diamond in its setting, are the four nitrogens of the four rings. The nitrogens hold an atom of iron in the case of the hemoglobin molecule. The nitrogens hold an atom of magnesium at the center of the chlorophyll.

Both of these metals, the iron and the magnesium, present two faces to the world. And these two faces operate like a quantum clock that tick-tocks with time. In one quantum state both metals are loaded with incoming electron energy. They tick into one valence. And in the other quantum state the metals unload and they tock into a second valence. Both metals, when held by nitrogen, tick-tock for all of their working life in a quantum state.

This is how oxygen is passed into the hemoglobin molecule that sits inside its doughnut sac. This is how oxygen gets into every tissue of the body to bathe it with its kiss of life. And this is how oxygen gets delivered from the chloroplast sacs in the mesophyll of the leaves of all plants and trees. Mesophyll is the tissue that opens up into the breathing pores or stomata of a leaf. This tissue is the flexible carrier bag for chloroplastic sacs. It is surrounded by air spaces. These air spaces are open to the atmosphere for gas exchange. This, too, is how oxygen is delivered into the atmosphere and to the oceans and the soils of the world for all of the aerobic life-forms within these systems. It seems like part of a divine plan, these twin sister molecules working hand-in-hand in their quantum homes to forge life for the entire planet.

So, what is old is new. Red and green have been mystical colors since ancient times. What is new is the story these colors hold for the human race, but like the old, their importance goes beyond human language and the written word into the mystical Celtic conundrum of the question of the meaning of wisdom itself.

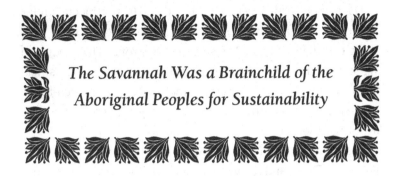

The Savannah Was a Brainchild of the Aboriginal Peoples for Sustainability

THE SWEET SAVANNAH

The voice of the aboriginal was heard in North America long before that of the pioneer. This voice was a sacred one on the landscape. It begat an oral tradition. The words that were spoken were not wasted. They arose afresh from thought into a tradition of dance and meditation. The silence of the continent distilled the words and focused the thoughts behind them. Each word was born in its own silence. Out of this embryo of silence came the Savannah.

The most daring idea in all of the world at that time was the Savannah. It was a concept of land governance that multiplied game and kept the landscape intact. As the architecture of the trees arose, it housed the wilderness at its feet. This stable monument nurtured a wildflower meadow that was grazed and relaid with the seasons. A fine tuning was weaned out of the wildflowers by burning. This adaptation to fire was felt by the trees, copied by seeds, and greeted by grasses. It made many plants of North America unique.

The face of North America is a long one. The continental profile catches the sun fresh from the east. The solar exposure to the land of the continent is high. The sun itself is strong. The product of this strength is fed into the North American trees. These trees yield crops that are immense. The crop is the nut for non-nitrogen-fixing trees. And the crop is the pod for the nitrogen fixers. These trees favor this landscape.

The native trees of the Savannah are the oaks, *Quercus,* the hickories, *Carya*, and the walnuts, *Juglans*. These trees spread out a mighty canopy to the sun in great arches of horizontal branches. These deciduous natives have bud tips that are clocked into spring temperatures. When the heat units are sufficient the apical meristems of growth unfurl their green flagship to the sun. The leaves, too, reach a rapid maturity for photon trapping. The leafstalks called petioles move, tracking the sun, and the midribs of the leaves are stretched and taut for the photon catch. The roots of these trees bore down into the soil and anchor the canopy in place.

These trees have developed their own quirks. The oaks produce their own sunscreen. It functions as a suntanning cream for these sunbathing trees. These aromatic biochemicals are called quercitrin and quercetin. They absorb the high-energy region of the light spectrum and resonate the excess energy in a quantum change. This bleeds off any damage this energy may cause to the internal metabolism of the tree itself. The hickories, *Carya* species, are masterminds of carbon sequestration. These trees feed the carbon into the organic chemicals that make up the dense polymeric structure of hickory woods. And the walnuts, *Juglans* species, produce root allelochemicals that act as a soil drench of chemistry, scouring the ground around the mother tree and keeping the area free of her children's saplings. This is because the competition would kill both mother and child. The pod-producing trees, like the honey locust, have mobile leaves that open and close. At night and on cloudy days the leaves are closed. They open flat to receive the sun.

The vast parklands of the Savannah were flash fired in April and November. At these times the native grasses are brown and dry. The roots nearer the ground are still damp and safe from fire. The woodland perennials, too, are asleep in dormancy. The fire spreads very rapidly with these light burnings but stops at the dry rough trunks of the trees. Many of these species, too, became fire resistant over the millennia. The flash firing produced a fine powdered ash. This ash condensed one important element for the Savannah. It was potash.

From the dried ash the potash seeks water. This becomes potassium hydroxide. This chemical is a fertilizer, especially for nut trees. Potassium hydroxide also protects cellular tissue against the oncoming frost and the crystalline damage of winter. It makes the potassium readily available for the forthcoming nut crop because it is water soluble. In addition, a potassium hydroxide solution is a potent fungicide. Landing near the trunks of the trees, it protects the trunks during the fall and winter months from fungal spores, especially the heavier ascospores. The ash creates a high pH on the surrounding plant surfaces and acts as an insecticide, keeping populations of pathogens low.

In the past, the trees of the Savannah were groomed by two species of birds. The larger, the heath hen, particularly liked the succulent curculios. These nut weevils arrived to feast and to lay eggs on the large and swelling buds of the spring. Once nipped, these sweetened buds were vulnerable to disease. A little later in the season carrier pigeons flocked in to feed on the insect populations of the Savannah trees. It has been said that the sun was darkened with the density of these incoming flocks. The heath hen and the carrier pigeon are now extinct. The hungry pioneers did not know of these birds' importance to the Savannah.

The Savannah produced enormous crops of nuts. These were a first-class protein, fat, and carbohydrate food for human consumption. It did not go unnoticed by the animal world. The squirrels, with their love of the nut sugar *d*-Quercitol, were first off the mark. Then in succession all of the mammals and large birds increased in number based on the nut food source and predation. The foxes, coyotes, and wolves increased. But the food that was the lifeline of the aboriginal peoples increased by a hundred fold. That was the deer herd. The deer represented food and clothing to the aboriginal peoples. It meant something far greater, it meant sustainability on a landscape that could be harsh to them and to their children. The Savannah created sustainability for food, for water, and for life itself.

The Savannah was a concept in American commonage that blended the tree with its habitat into a canopied parkland for successful hunting. It was a brainchild of the native peoples on this continent, an idea so singular that it was immediately copied. It can be seen today in splendor, around Europe's castles of the powerful, the rich, and the famous.

MAGICAL TREES

 Magical trees and mystical forests are as old as father time. Magic is known and recognized worldwide both in the past and in the present. The word itself is very old and arises with the birth of civilization. Its source is from the Old Persian *maguš,* meaning sorcerer. And magic is universally understood as something that has extraordinary powers bordering on the supernatural. Some trees of the global garden fit the bill and these trees are indeed magical trees.

The elderberry, *Sambucus,* and the hawthorn, *Crataegus,* have long been considered to hold magic. This knowledge was held parallel in time by different cultures of the new and old worlds. It was known in China and Japan as well as Russia.

The elderberry was a favorite in the heady days of the Egyptians and has remained in continuous use to modern times. In Northern Europe many people would never pass an elderberry without greeting the tree, either in word or deed. They would doff their hat in respect and bow low. This species was never

15

burned or destroyed, because it was believed that the souls who lived in the elderberry would then be destroyed too. It did not go unnoticed that the elderberry is never touched by insects or destroyed by disease.

But the elderberry does have extraordinary powers. In ancient Egypt it was used as a valuable cosmetic. Indeed, its effects on the skin are rejuvenation and revitalization. The flowers were popular as an infusion used as an eye lotion to rejuvenate weary eyes.

The Seneca aboriginal peoples used the elderberry on their premature and newborn babies. These babies were washed in a lukewarm water in which dried elderberry flowers had been previously soaked. The warm waters contain capillary-protective biochemicals that act as a gentle skin tonic through vasodilation on the newborn skin.

Probably one of the most important biochemicals is in the mature black fruit of the elderberry. It is a rhamnose sugar complex. It increases the efficiency of the metabolism in the avian eye, aiding visual acuity in the transition zones from darkness to light and vice-versa. This sugar is much sought after by songbirds in their north-south migrations of the global garden.

One tree that has held the world in its magical grip from the Americas to Europe and into ancient China was the hawthorn, *Crataegus*. In rural Ireland and China, superstitions still abound about the hawthorn. No child is allowed to bring a flowering branch of hawthorn into the house in May. It was said that the flowers brought bad luck to the house for one year. And marriage proposals were often accompanied by a blackthorn walking stick for good luck. These sticks are now a tourist industry. In China the hawthorn fruit was flattened and de-nutted. The fruit was mounted on a stick. This popsicle of fruit was given to children as a natural health food.

The hawthorn, *Crataegus*, is a member of the *Rosaceae* family. It is closely related to the apple. The fruit of the hawthorn is a pomme and it has five pips or nutlets inside. The fruit is edible, especially so after the first killing frosts when the sugar complexes change into a sweeter form. Sometimes the skins have russeting. These pommes are the sweetest of the sweet. The nutlets were collected by the aboriginal peoples, stored and dried. They were ground and made into a coffee. Like all regular coffee, they have high concentrations of caffeine.

In the global garden the hawthorn has an effect on all butterfly populations that improves their health and their ability to survive the ordeal of migration. The mature leaf produces a hormone that has a direct impact on the growth and development of the caterpillar. The hormone is a strength enhancer. In addition, the leaf holds biochemicals that make high-energy ATP, adenosine triphosphate, which is the fission fuel firing the energy system for migration itself.

The greatest magic of the hawthorn is a unique biochemical commercially known as Curtacrat or Crataegus-Kreussler, which affects the beating heart. In the heart there is an artery called the left ascending coronary artery. This little strip of plumbing feeds the muscles of the heart itself with much-needed oxygen. This is the pipeline that can get blocked. And this is the region that requires the now ubiquitous bypass surgery when it gets severely blocked. The biochemical of the hawthorn works on this little piece of plumbing, opening it and keeping it open. This cardiotonic is said to have a hypotensive action and function on the heart. It also has a beneficial effect on Stage II congestive heart failure, by opening up that most important feeding pipeline of the heart, the coronary artery.

In the farming community of North America, there arose a custom borrowed from the aboriginals around the hawthorn. Of a summer's evening, while homeward bound with his dairy herd in front of him, the farmer would pause to pick a trail snack of sweet red hawthorn pommes. He would munch the fruit and spit out the nutlets. This simple little act kept the farmer's heart healthy.

There is also another magical tree, which is coming into the modern world from an older culture. This is a rain forest tree of a small Samoan island in the southern Pacific Ocean. It is *Homolanthus nutans*, an aspenlike tree. This tree is more recent. The medicine men of the Samoan island have been using this tree in the treatment of yellow fever and hepatitis for a long time. Now it represents a new frontier in the management of the HIV virus. An antiviral biochemical, prostratin, has been extracted from the living plant tissue. It blocks the virus from replication just as an antibiotic blocks a bacterium. The trees of the rain forest have spoken again and there is more than a mote of magic in them, this time, right enough.

The Global Forest Has Within Itself a Master Plan for Sustainability

A SUIT FOR SUSTAINABILITY

The forest can outpace the brains of Wall Street for mutual funds. Bankers have bought forests. Japan, with its linear market, has a tight grip on its own forests. The Japanese will not sell their own precious primary reserves of timber. They prefer their forests intact. Within North America's forests there are trees of extraordinary value. But nobody has had the thought to grow them as a financial cushion for sustainable living.

Not since the Middle Ages has the Western forest been used intelligently. In medieval England the primary resources of the coppiced forest produced sixty-four items of market value, such as charcoal for fuel, rods and twigs for wattle-and-daub construction, and long poles used to support hop vines. Many of these were used locally, but others were sold for an expanding global trade. The English forests were carefully cut on a seven-year renewal cycle that was of itself sustainable in a mild, almost frost-free climate. Everybody benefited, the rich more than the poor, of course.

On the North American continent a similar concept of sustain-ability in perpetuity existed which rose from a different practice based again on climate. The increased sun and solar exposure pro-duced a bounty of crops from the Savannah's sustainable design. The tree crop fed the continental wildlife and the aboriginal peo-ples who depended on this food. The Savannah, too, was used for the common good and everybody benefited in this paradigm.

Now, the human heritage and birthright of the global forests are owned in gross default by faceless names and nameless faces. Nobody benefits.

Out of the flux of today with billions fleeing into the famine of urbanization, the small farmer can rise again with a bioplan in hand. A small farmer carries a vocation in his heart. It is possibly the most important vocation on the planet. It is the gift to grow. It is the knowledge and wisdom to take the seed from youth to adolescence through to maturity. This gift brings its own reward in crops. But the crop can also be trees. The trees can be the cash crop for the farmer in a bioplan, whether his farm be large or small.

A bioplan is sustainable and increases diversity for predation and prey; a free natural management of a natural system. A bio-plan will walk organic farming one step further to increase the biodiversity of native species of plants and animals. Quite often an organic farm, good though it may be, can be a desert too, if the farm is just composed of mile upon mile of crops in empty acreage. The forest must come back to the farm in the form of an orchard, nut orchards, and set-asides of select trees. Beneficial trees can be added to existing hedgerows or fencerows as anchor trees. New hedgerows can be added as windbreaks to prevent tur-bulence or wind damage to crops. Many cereal crops don't have sufficient tillering or buttressing for the strong winds of global warming. These trees also stop erosion of the surface soil which is the farmer's bloodline.

The native trees of a bioplan grow exponentially. That is their advantage. They are true perennials and once planted they will grow on. Trees can be planted on any farm, in almost every temperate zone in which farming takes place, in either large or small set-asides. These areas of growth within the farm amplify biodiversity, purify groundwater, control pollution, decrease nitrate pollution, and multiply songbird populations, mineral recycling, and wind buffering. Soaring temperatures and solar ultraviolet radiation are tamped down. There is an increase in grace and beauty. Trees are the last items of consideration when they should be the first.

Examples of beneficial trees for the farm's bioplan would be any native species from the huge and global *Rosaceae* or rose family such as apple and cherry trees, which are valuable because of the shape or morphology of the flowers. They are saucer-shaped. Pollinating insects can rotate in flight inside the flower to gather nectars and pollen. They cross-pollinate different flowers of similar species, improving the production and quality of the fruit. Also beneficial are those from the *Leguminosae* or pea family, which includes the honey locusts and laburnums, whose natural nitrogen-fixing ability is invaluable to the farm. Beneficial insects and pollinators such as honeybees, ichneumonid wasps, chalcids, and native bees use these flowers because they can rotate inside of their form and feed on nectar, pollen, and honeydew, all of which are produced in very large amounts.

The *Leguminosae* offers food to a different kind of flying beneficial insect. The native pollinators like bumblebees that are heavier and stronger in build take on the legume flower, and walk to the staminal tube. Both pollen and nectar from these flowers feed another team of players like the green metallic cuckoo wasps and leaf cutter bees, who are important in field pollination and in predation itself.

Growing cash crops of set-asides of trees is financial planning for the future. Farms and farmers desperately need this planning. *Juglans nigra*, black walnut, is one such tree. Walnuts are grown for their chocolate-colored wood, which is stable as veneer. Dark luster woods are rare in the global market and are much in demand. A single straight bole of walnut that can be used for veneer can bring in $60,000 at auction. This will go a long way to paying for a university education of a child or grandchild.

In addition, in both North and South America the *Carya* or hickory species can be grown for their nut crops. There is a lively market for nut meats, nut shells, and nut flours. Nuts are also used in the alcohol industry in nut liqueurs. Set-asides of the native species of *Gleditsia*, honey locust, in North America, Africa, and Asia would considerably add value to any farm. At present, the powerful wood of the honey locust represents a four-billion-dollar business in North America. This wood is termite- and insect-proof. It is an excellent substitute for pressure-treated woods. The pod crop from honey locust is a source of flour, of dairy ration when the dried pods are fed with hay, and various medicines. It is also a tree used by the hive.

Trees like *Asimina*, pawpaw, can be grown for their delicious pineapple-mango-banana-tasting fruit, both in North America and tropical areas. They can be coppiced for medicines called acetogenins to treat drug-resistant cancers, and for naturally occurring, biologically degradable, all-purpose insect repellents. The pawpaw is also a butterfly tree.

The wine industry in North and South America, in France, England, Germany, and Australia, all need a supply of oak for aging wine in casks. A similar need exists for the whiskey industry. There is little supply and great demand for *Quercus*, oak. The North American continent grows the truest and best oak. It is a specialty of the native forests. Many of the species of oak produce woods that are unique to this continent. Oak is an ideal set-aside for a future cash crop.

The small family farm can reach for sustainability in both land and financial management by an imaginative approach to the future. The answers are out there, in the forest, where they always have been . . . waiting.

The Trees and the Forests of the World Exist in God

THE PARANORMAL

The early morning of eastern Canada is filled with a golden light. It happens all too quickly, a blink after sunrise when the air itself lifts, in a delicate surprise, at its own splendor. There is a point at which this first light lands on the trees and branches of the forest and they become slashed with ribbons of scarlet. These red ribs of light rope in each tree, brindled, for another sultry day.

And so . . . like the Queen of England, the farm woman rose to eat her brown egg. It had been laid the day before. The white had curdled a little with cooking and the yolk was soft. Perfect. A piece of homemade bread was washed down with a cup of tea. Irish, brewed with well water. The leather work gloves were somewhere in the kitchen. They were small. The hand was feminine.

She walked out of the kitchen door, closing it against the lullaby of a little life. She quickly passed the perennial border to her left, glancing at it while she pulled on each glove. The flowers were opening their petals in a yawn of fragrance, getting ready for the business of bees. A stray moth moved like a mote of pleasure. The spiderwebs gleamed with tiny gems of crystalline water in catchment caves near the rich earth. The line of espaliered apples pointed the way to the trees of the forest.

Then she passed the chicken house. The females were clucking among themselves about another egg. She could hear a thump as one fat hen landed in a nesting box. First egg of the day. She could hear the murmur of mature hens muttering the mumble of the fowl world. The feeder rattled, ringing the chain. And all was silent as the flock poised to listen to her soft footsteps pass them by. Then, as she was almost into the trees, she could hear a clucking begin to break into a cacophony of curiosity.

She could smell the soft fragrance of the cedars almost before she set foot into the forest. An August sun had droned through the evergreens the day before, opening up the glands in the leaves. The thin soil was filled with roots bulging up at the edge of the forest and she had to step over them. The mosses were dulled with misery; their spore children had fled into the winds. One lone *Oenothera* flowered, crouching against the soil, the yellow sharp with a brittleness of acidic color.

She walked into the forest following the pathway of the day before. She kept going until she arrived at the grass-filled clearing. The trunks of the trees were in one great circle around her with their canopy closing in the sky overhead. Some trunks were fat and squat, others were more slender with youth. These leaned together or apart in a conversational way. The trees formed a palisade around her. All she could see was this assortment of trunks.

She stopped in the center of the space. She was content to wait. Her husband would not be long now. The trees held a silence; not a whisper of a breeze could be felt. The air itself was still. A wall of peace approached her from the trees, gently moving in on her. She did not notice. The wall edged in closer. Then it touched her. She became aware of it and its absorbing presence.

Like a soft, sweet, summer rain she became submerged in the stillness of the place. She was aware that it was all around her and it was coming from the trees because it was a part of them. She could feel herself breathing more slowly and more deeply with their gentle rhythm of relaxation. She did not move. She breathed in this peace. It manned her inner being, taking her over entirely. Time stopped and moved from somewhere to nowhere.

Prayer started to leaven inside her. It rose up into her mouth. Softly she began the Gaelic prayers of her childhood, spilling them out into the forest around her. The prayers themselves became one with the trees of the forest. Outside of them stretched eternity, an infinite world of unity in which everything was the present time and held in that special form of order that went on for ever and ever in all the directions of reality. Lulled by such peace she leaned into it. Her body moved forward a touch. The prayer and its meditation filled her up. She was her own vessel. She became adjunct to her own peace. It was the peace of the universe. This peace was smooth and it was soft. It fed itself into all spaces. It was in the trees. It was of the trees, the trees of the forest.

Suddenly she became conscious of something else transpiring. This had been going on, for how long, she did not know, but she became aware of it with a new reality, a new understanding that surprised her. As she had leaned her body toward the trees in the full lull of peace, the trees, all of them, had lured themselves too, toward her, in a conspiracy of one another. As she prayed and meditated into their combined peace, the trees had done the same. They had been leaning their trunks toward her also. She realized with a start that the trees were praying, too. The trees and the forest were praying too because both of them shared the same God.

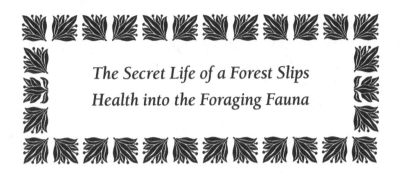

The Secret Life of a Forest Slips
Health into the Foraging Fauna

HOME, SWEET HOME

A forest is a home. All the forests of the global garden are homes to microbiota, insects, birds, mammals, and plants. These homes are important to each and every form of life. No one species is better or worse than the other. They are equal to one another in a chain of connectivity. Each bee, each wolf has the right to dream or die, has the right to live a life, its own particular life, of wonder. And it has a right to that home until the end of time.

In the forest there is a rule of thumb that amplifies diversity. Each species of tree is responsible for about forty species of insect. These insects are associated with the lifestyle of that particular tree species. A diverse forest therefore spells biodiversity, explodes and amplifies diversity in every range possible from the visible to the invisible. It sets a pattern for predation and prey. It lays the foundations of health.

Every forest has an invisible life. This powerful life force is found within each tree and is seen only in the death of the tree itself. In the forest soil around the dying tree the invisible endophytic fungi are turned into the visible as a carpet of mushrooms. These fungi had been living in an asexual way within the tree around the darkness and dampness of its plumbing system. The ribbons of hyphae of the fungus, trafficking chemistry, work for their consumer needs. Some of the most potent medicines in the world come from these hidden hyphae homes. Many are used in the treatment and management of female cancers, the class of taxanes being the most prominent in present times.

The fungal species that set up house within the anatomy of a tree are somewhat similar to the chip of mitochondrial DNA that is gender specific and is inherited from mother to daughter in the long line of human life. The endophytic fungi, too, are tree species specific. The morel mushroom nests with the American elm and the giant puffball with the standard apple. The endophytic fungi may be one of the principal regulators of carbon exchange. They may represent a cycle of life similar to that seen in the algae, mosses, ferns, and gymnosperms, where an asexual phase is followed by a sexual one. For the algae of the great oceans of the planet, the asexual phase may last a few days, but for the older trees, this time is counted in hundreds of years, sometimes running into millennia.

In North American forests, squirrels are nature's primary foresters. They crave a form of sugar, *d*-Quercitol, that is found in healthy nuts. They bury many of these nuts and a forest is born or regenerated from the best of the best. Where there is a generous squirrel population within a forest all the larger mammals move in, from fishers and foxes to bobcats, wolves, and coyotes. These are fed from a baseline of rodents who also enjoy the fruits of the forest stolen from hidden squirrel hoards. Rabbits and hares have an ebb and flow with predation from other mammals and raptors.

The larger predators like cats and bears spread out in their territory for hunting. In between all of this the deer numbers climb, feeding on acorn mast. They, too, will have their day with coyotes and wolves.

There is mass transit in the global garden. The seasonal exodus from south to north and back again is common for songbirds, many species of butterflies, and other insects. The long slog of caribou and reindeer and their eternal search for hidden lichens carries them across the north. For much of the fauna of the world the forest represents staging grounds in these migrations. The food that is on offer for migration comes from the nectaries on leaves and flowers. Protein, too, comes from the vast array of forest pollens and complex sugars found in extra floral nectaries, in the bract openings of flowers, and in the sweet seals of the tips of most buds. Quite often feeding is itself part of a larger and sometimes unnoticed cycle within the forest. This cycle connects forest with forest and continent with continent as if the world were one.

All trees of the forest produce sap. Many have a sweeter sap than others. These trees are closely watched by squirrels that remove the bark down to the cambial layer. These tissues then exude a sugar solution that collects. If the air is cold, this will become a popsicle stalactite. This is eaten by the squirrel. Then a winter bird, the chickadee, drinks the sugary exudate. A little later butterflies like the morning cloak will feed. The ants are next, and the wound calluses over. But this nip and tuck of the cambian layer makes a tree bear more fruit. And the cycle of feeding replenishes itself.

The fauna, in turn, repay the forest for food. They help in the dehiscence or opening of seeds. Sometimes the seeds are like Velcro and stick to the animal's coat; other times they are fine and will track with the mammals until the next rub or shake. But more often the seed needs to go through an acid bath to scarify the tough outer pericarp of the seed's coat. The acid bath is the hydrochloric acid in the mammal's stomach. The seed slips through the large bowel and is deposited ripe for growing. In the case of the pawpaw, the seed must be fondled by a raccoon, which imparts a layer of hand grease on the seed. This seals the seed, making the testa airtight. This prevents infection of the embryo and the seed is ready for its jump into elongation and life.

Many beneficial insects use deciduous leaves for overwintering purposes. Lady beetles congregate into large overwintering colonies. They do this in the Muir Woods of California and they do it in the eastern seaboard forests in New England. These insects use the black box warming effect of leaves, which behave as small pup tents for these beneficials to survive and communicate in great numbers. As soon as the temperatures increase they are off in the race of forest insect predation.

Forests are also holistic in their view of fish. Many riparian forests produce biochemicals that have a benign effect on water systems. The *Juglandaceae* or walnut family drops juglone sedatives into the water in the fall. These sedatives affect the dormancy of fish and water creatures, helping to stabilize and maintain their lowered basal metabolic rate. The maritime forests tamp down nutrient flow and decrease toxic algal blooms. Forests spread a hand of blessing over the creatures in their care. This blessing is called health. And no doubt, it is for us all.

The Broken Forest Is in
Our Children's Tears

HEROES AND HORMONES

The legends of heroes climb out of the past. They are part of every ethnic group. They molded every culture into being. A hero is plucked out of life and put at the top of the pecking order. There they stay to be admired and loved. Some forever.

All cultures have had their heroes. Once upon a time in Ireland it was a blond leader of men, Fionn and the *filíochta*. The "Fair One" and his band of warrior-poets ran the length and breadth of the island of Ireland or Éire. Their giant Irish wolfhounds loped by their sides to protect the little green island from invasions. The memory of Fionn and his warrior-poets lies deep in the heart of Irish mythology, poetry, and memory. Indeed, it exists for us today in the contests of the Olympiad. The winners of the Olympic games who once upon a time received their evergreen crowns of fresh olives, *Olea europaea*, now get a precious metal of bronze, silver, or gold.

The aboriginal peoples loved youth next to wisdom in their heroes, as every male child realized when his childhood began to wind down. The ritual of puberty involved a pumping of the young male with anabolic hormones prior to his trial of endurance. The youth passed into manhood enduring a solitary life in the open wilderness and feeding himself with his wit and skill. The rite involved a body building plant hormone from the native medicinal plant of Culver's root, *Veronicastrum virginicum*.

Heroes and hormones go together. They have a common face, that of prowess and body power. Both of these rest on hormones, both male and female hormones. It was the story of the heroes of the past. It is the story of the heroes of today, too. The banner of the sexes is seen on a daily basis, creaming off the milk of the media with a tantalizing taste to remember.

The hormones that build the body and the human face also exist elsewhere in nature. Hormones are found in all plants and trees. Plant hormones and animal hormones are alike. They have a similar molecular biology. It is as if one is the mirror of the other in some magic mirror show . . . "And who is the fairest of us all?"

Heroes and hormones are tied into body image. This is in turn bound up with sexual success and reproduction. The body is the advertising for the hormone in a subliminal message of sex. The chemical hook of it hangs in the mind like a light bulb that can go on and off. This creates an eternal interest and excitement in the subject of sex.

Plants and trees have successful sex lives, too. Trees have a similar hormonal identity to mammals. They, too, must get big and strong prior to the vigors of sexual success. This, for some trees, may take up to a century. Many specialized plants may take that long, too, such as the bizarre Welwitschia family of gymnosperms of the deserts of southwestern Africa, more than a few species of new world cacti, and of course, the woolly-haired Dicksonia ferns. Then after the initiation of sex, trees are programmed to live for a long time to spin out their sexual lives. This is seen in the amassed tissue called wood which is loaded with hormones not too dissimilar in chemistry to the mammals. In fact, if the tree's hormones are examined in a mirror, they look quite similar to the mammals' counterparts. This is not surprising because both are built from almost identical aromatic hydrocarbons. They are called isomers. These isomeric compounds have strange spacial relationships to one another of identity; one chemical spells the reverse of the other.

The hormones for sex in a tree are called gibberellins. There are others also. The gibberellin is of a family of related biochemicals that are expressed in the spring at the tips of branches. These chemicals prepare and groom the tips for flowering and later on in the season for fruiting. The phenomenon of fruiting is the cost-effective coupling of the tree in its efforts to produce embryos in seeds. These seeds are the next generation of trees. Like mammals, trees mother and nurse their seeds and seedlings in every possible protective way.

In the past two hundred years the North American forests have been cut down. Many of these trees have been pulped for paper and have been milled for their wooden products. All of these activities have taken place near large bodies of water. This is because part of the processing of the lumber of trees needs water either for storage or for transport. The best of the best of the native forests have been cut. These trees were the healthiest and were the greatest producers of seeds. So these trees, too, carried the greatest hormonal load. These hormones are water soluble.

Plant hormones from the forests are now found in drinking water. In addition to these hormones, fresh water also carries a new burden of birth control pills and other modern hormonal medicines. These have been excreted as excess chemicals in urine. They are from manufactured sources. This, in turn, makes a cocktail with many related pesticides. These mix and flow in volume. They act synergistically with themselves and with one another. This hearty freshwater soup is the potable drinking water of North America. The foreign hormones are called xenochemicals. These are physiologically active sex hormones, all ripe to run the race of the sexual game. They are now found in the bodies of all mammals, including the human body, as a new molecular pollution.

The hormones that were once in the delicate balance of nature within the forests are now uncontrolled in our bodies. They are in one-third of all human embryos and the amount of xenochemical found is about equal to the in vitro, naturally produced hormone levels. Our broken forest is in our hearts and in our children's tears. . . .

The Nut Trees of North America
Were Called Antifamine Trees

A HANDFUL OF NUTS

There are nuts to die for in North America. They have been part of the native wildwood forests for a very long time. In the hungry past, nut trees were known to the aboriginal peoples as antifamine trees. This was because famine stalked this continent in climatic cycles. The edible landscape kept the aboriginal peoples alive throughout the bad times of collective hunger. So the location of nut trees was always noted. This knowledge was passed from one generation to another as survival information.

Nuts come in all shapes and sizes, and many plant families produce them. These families have representatives across the world in other regions of the global forest. The nuts of related families are similar in shape but are rarely identical in taste because of the variation in local soils, which changes the trees' biochemistry.

A nut is a fruit. It is most often single-seeded. The seed is protected by a pericarp coat which, depending on the family line, might be woodlike or leatherlike. This in turn is usually protected by a husk. As the nut matures, the husk may open but the pericarp coat never does. The seal is forever, except when acted on by the enzymes of germination.

The delicate food-filled embryo inside is protected by the strength of the pericarp wall. This wall also delineates fertility within the embryo. If the embryonic center dries down a little, the space formed is a trigger for germination within the nut. The enzyme and osmotic forces in germination and subsequent growth are powerful enough to rend apart the pericarp into two fragments. The physics of this natural act are extraordinarily powerful.

Many North American nuts have a husk or a partial husk similar to the filigreed cup of the oak species. The husk itself is a common enough sight, being green at maturity and rapidly turning black or brown shortly after hitting the ground. The husk also has a microscopic surface landscape of ridges and folds. In the case of the black walnut, *Juglans nigra,* and the butternut, *J. cinerea*, these ridges have additional features—glandular hairs carrying an arsenal of explosive chemistry. Some of these chemicals are iodine-based and release a pungent smell of an iodine aerosol into the surrounding air. These are protective biochemicals. Simply by holding a green walnut, *J. nigra*, a young child will receive protection from early childhood leukemia. But despite their common face, all husks of all North American nut species have been ignored by science.

The vast stands of American chestnut, *Castanea dentata*, of the beech family, are still the most famous of the North American continent. The tragedy of a blight, a canker-forming bark disease, *Endothea parasitica*, first discovered in New York City in 1904, added to their legend. The dark shiny nuts encased by a hedgehog husk were one of the special treats of the continent. They could be eaten fresh or exploded into flour by heat. The central flat nut of a triplet set squashed into the middle was always the very best. The chestnut is making a comeback. Time has favored the genetic code with resistance. The tree is relearning the lessons of life. It will be on the stage again soon. A related nut is the Sierra chinquapin, *Castanopsis sempervirens.* Less than nothing is known about this small but important nut, which defies transplanting. For this reason alone it is an important feral gene in the diminishing gene pool of the beech family of the Americas.

The smaller and sweeter nuts of the North American West are found in the pine, *Pinus,* family. Some of the larger ones are of the western digger pine, *P. sabiniana.* These are about the size of a broad bean. Pine nuts from the single-leaf piñon pine, *P. monophylla,* have a soft shell and nuts from the two-leaf piñon pine, *P. edulis*, add a butter flavor to the nutmeat. The rights to harvest these wild crops of pine nuts are sold at auction, and potential pickers bid against one another for this eight-million-pound harvest.

The favorite and most delectable nuts in North America are the butter-flavored pecans, *Carya illinoensis.* These large nuts have a very extensive market. A close second are the shagbarks, *C. ovata*, and kingnuts, *C. laciniosa.* These are now available freshly picked and dehusked in open-air farmers' markets. Their popularity is rising. The nuts of the *Juglandaceae,* or walnut family, are enjoying a resurgence. These are the black walnut, *J. nigra*, and the butternut, *J. cinerea.* The meats of these walnuts are unique. They can undergo the process of industrial food manufacture into various popular foods without a change in flavor; black walnut ice cream is an example. Then, when the nutmeats are extracted, the woodlike pericarp shells are used in the pressure washing of the façades of older buildings and the sandblasting of paint on aircraft. The pericarps scrape away the detritus of pollution and paint like large-grained sandpaper without harming the under-lying layers.

Probably the most important nut to the aboriginal peoples for everyday use was the bur oak, *Quercus macrocarpa*. This large-seeded oak produces a yearly crop of very sweet edible nuts. The flavor of these nuts also varies from location to location, the sweeter being found in karst or limestone areas. The bur oak acorns were collected in great numbers. These were bagged in handwoven sacs made from basswood, *Tilia americana*, fibers. They were stored over winter in running streams.

The acorn nuts were eaten as a vegetable. They were cooked over an open fire and the pericarp peeled. The meat was also dried and ground into flour. A bannock, a kind of sweet scone, was baked with the flour. It was said to have had the flavor of mild peanuts and a rich rose color. These were cooked without salt, of course. Apart from using bloodred samphire, *Salicornia rubra*, in the boreal north, table salt or sodium chloride was not common on this continent. Wild herbs were used instead to enhance the flavors of food. Times have changed.

The Forest and the Fairy Form the
Landscape of a Child's Imagination

THE FOREST, THE FAIRY, AND THE CHILD

There is so much to a child. Children are born with knowledge. Each conception confers that knowledge with pairing. It is found in the print of DNA. All that was known before and all that will be needed is fed into that new molecule for this emerging life. Contrary to general opinion that no possessions are carried with us into the next life, all of us do carry something. We leave this life with the full gift of our knowledge of birth together with the wisdom garnered from it, through the portal of the grave to beyond.

In the sunshine of youth, children will express their feelings through art. Drawing is always in the surface landscape of the mind ready to leap into form, on a floor, under a table, on a wall, or even into a valuable book. Stick figures emerge first on paper. These images are powerful to a child. Their cartoons hold the birth of self. Often there is another, a sibling, a sun, a family pet. And almost always a house with a chimney producing smoke. Then there is the tree. This is the expression of the important items to a child. The drawing is given only as a gift. A concession of the child to the adult world for full understanding. Because that drawing expresses all that is important to the growing child, such gifts are serious.

Because children live outside of the box, they are still trapped in their original wisdom and thought. Culture and civilization have not yet whitewashed the patterns of their mind. That mind takes flight while the body is still anchored in reality. The flight is guided by instinct, through which clarity is absolute. A young child can read the language of a baby with telepathy, as they can distinguish male from female in the very young as if by some unerring instinct. A young child can connect with the mind of its mother to initiate the flow of milk before the hunger cry. They can enjoy the companionship of invisible people. And . . . then a young child has an innate understanding of trees.

To cut down a tree is murder to a child. The tree is snapped out of the drawing and its life is lost. A child knows that a tree has a life. To take down a tree is an unthinkable act of savagery to a child, because a tree is a friend.

The tree is an object solidly anchored in the landscape of identity, which in turn becomes the landscape of the mind. This is a touchstone to which the self returns for reassurance throughout life. It anchors sanity and healing together in a spiritual life of the mind. Such anchors established in youth serve throughout life to stabilize the spirit and produce the wisdom of old age.

To cut down the global forest is sheer madness to the mind of a child. It is a senseless act of willful destruction to take the lives of the trees of the forest. It offends the songbirds whose nesting will disappear. The food for the birds is taken away, too. The butterflies have nowhere to be beautiful, to show their woven colors on silken wings. The great animals of the forest will be lost. The lesser animals have nowhere to hide, to sleep and polish their whiskers. The slugs and snails will move their homes in endless tracing. There will be no trees to climb or swing under in the long timelessness of childhood.

If a child believes in trees, then they believe in fairies, too. For sometimes these fairies live in trees. Fairies can fly too. They can travel around the branches of trees and they can sit on a particularly large leaf just as an angel or many angels can sit on the head of a pin. It is always that easy. It says so in fairy tales, in books, in stories, and most of all in bedtime stories all about the fairy world.

The fairy can sometimes wear flowers in her hair. These blooms have to be very tiny. They are tied to curls with satin ribbons. The fairy has clothes too. They are usually cut out of leaves. Sometimes they are cut out of the dense petals of flowers. The scallops make the bottom of her skirt. She always has bare legs and on her feet she wears shoes. The shoes are golden. They are slipped on as a new pair any time she wants when she passes a gathering of lady's slipper orchids. The fairy dips them into the golden light of dawn for coloring. She must carry a wand. It is a magical wand with untold power. There is a trumpet flower on the top of the wand. It is white and very fragrant. This flower is a cup from which her personal assistant, a green tree frog, drinks. He follows the fairy very slowly and carefully.

For some wonderful reason fairies have been held captive by children for over five thousand years, ever since the times of the *crannógs*, the ring forts of lakes, and the *ráth* or *dún*, the dry land forts of the open landscape. These Bronze Age sites of homes, especially in Ireland, carry their own taboos of behavior rich in Irish language, poetry, and ancient wisdom which are still obeyed to this day. They are called fairy circles, or the *Sí*, where the fairy people live.

Children are people. They are filled with dignity. They are little people who love and live too. They and their fairy folk will inherit this planet. They look to the adult generation ahead of them with confidence that they will be protected and cared for. They must believe that the pictorial landscape drawn by them as children is cared for, too. This is called human inheritance. It is a belief system on which our cultures rest and have rested to mold the societies in which we live. To cut down the global forest is a deep and personal betrayal of every child on this planet. It is a robbery of their imagination and a looting of their future.

For you see, there is really so much in a child. The conception of a child is the conception of all knowledge. Take away the tree and the fairy and you remove the child. This is the future. Listen to the child and remember the fairy . . . too.

The Trees Tell the Story
of the Americas

THE ABORIGINAL

History has swept across the face of the Americas and little is left behind. The trees tell a story of North America, for only they can speak of the cultures of the past. The cultures were oral cultures based on the traditions of the spoken word. Only silence remains. This silent trail is seen all over North America.

There were complex cultures in the Americas; some are known and others unknown. Some are discovered in surprising, out-of-the-way places, like Los Ajos in Uruguay, whose cultural passing was forced by the weather patterns of drought that turned rain forests into grasslands. These changes, too, marked the peoples of the Brazilian rain forest some four thousand years ago. But nothing entertains the mind like the medieval mound culture found in Cahokia, where servants were interred with their dead masters whether they liked it or not.

This mound culture arose well before the eleventh century on the greatest watershed of the continent on the river called the Mississippi. The present-day location is near East St. Louis, Illinois. It

44

was an extraordinary metropolis we have called the Monk's Mound of Cahokia. The city precinct gobbled an area of 300 acres. This is about 121 hectares in size. The total outlying area or urbanization was about 1,200 acres or 486 hectares, the size of a principality or walled city, European style. The population swelled to at least seventy thousand people. Both metropolis and population rivaled London, England, at that time.

The American peoples of Cahokia built mounds. These mounds were extraordinary in both mass and size. Indeed they rivaled another culture for their size, people who were gifted in building the extra large. These were the pharaohs of Egypt and their buildings were one of the wonders of the ancient world, the pyramids. The earthen mounds of Cahokia were greater than the pyramids, the largest one covering 16 acres (6.5 hectares) in the basal area and rising to 100 feet (30 meters) tall.

The secret to the well-being of the people who lived and loved in Cahokia was their own native black walnut, *Juglans nigra*. This nut-bearing tree was and still is the king of the forests of America. The tree stands out by itself in the flora of the forest. The apical meristems and the dormant tips have the chubby profile of the well fed. The art deco branches bear down to a trunk filled with sucrose sweet sap. The statement of the tree is that of command.

Even the litter of long leaves has a whispering harmony of release. All the leaves fall off the tree in one single day after a frost. And then the war chest of unparalleled protein is exposed. Booty for all.

A mature walnut will fit in the hand. It greens to maturity on the tree. Some trees produce an oval nut the shape of a hen's egg. These nuts are sweeter. Others are a perfect sphere. They are rough grained and hang either singly or in bunches of two or three. The nut of the walnut never varies in its packaging. There is an outside soft husk, green of course, which hardens and dries down to a deep brown when aged. Inside the husk lies a tough hard shell, wrinkled and worried looking. This cradles the nutmeat inside. The meat has the appearance of a human brain and it is closely encased by a thin vest or intima, a lacelike brown tissue. Black walnut meat has a unique ratio of first-class protein and essential fatty acids. Pound for pound, the nutmeat rivals a standing rib roast in nutrition.

It was the quality of this nutmeat that fueled Cahokia, that powered the muscles that built the earthen mound to rival the pyramids of Egypt. It was the medicine in the nutmeat that gave the people and the immense river their health. It was the fiber and its calcium in the nutmeat that anchored their collective insulin against diabetes. It was the ellagic acid complex that protected them from cancer. And finally it was the three essential fats—oleic, linoleic, and linolenic acids—that gave them the brainpower to do it.

The walnut forests that grew around the Monk's Mound of Cahokia in the deep alluvial river bottom soils of the Mississippi had something singular in their day. They were attended to and groomed by flocks of large chickenlike birds called heath hens, *Tympanuchus cupido cupido*. These hens would fly up into the massive branches of the walnut and stay there while they groomed the tree free of its curculio and other insect infestations. The chubby apical tips and sweet sap of the walnut were a constant lure for predation which the hens took care of with their razor-sharp beaks and lightning speed. The eyes of these hens could see the smallest snack. But these teams of pesticidal patrol are no longer with us. Unfortunately, they are extinct.

Alas, the passenger pigeon, *Ectopistes migratorius*, is also extinct. These migratory birds darkened the sun in clouds of movement. They groomed the walnuts of eastern North America for early spring egg casings and the smaller insect delights the great canopy offered. These, gentlest of birds, are now no longer with us.

In the end, it was the pioneer women who saved the last stand of virgin walnut forest. These fair ladies went by the hefty name of the Indiana Pioneer Mothers. They called their important bequest of *Juglans nigra*, black walnut, virgin forest, the Indiana Pioneer Mothers Memorial Forest. It lies in southern Indiana, south of Paoli. This bequest to the United States and to the world was a small parcel of sixty acres (twenty-five hectares) in total. It is a forest of living American history. It represents a history that once upon a time swept across the face of America and rivaled the great civilizations of the world. It is now brushed under a sixty-acre carpet of walnuts.

The Bioplan Will
Reverse Pollution

BIOPLAN FOR BIODIVERSITY

 In recent decades the human family has suffered from a fractured wisdom. The communal kitchen table is no longer used for eating. Grandmothers and grandfathers are caged. Their stories remain untold. Grandchildren do not benefit. Divinity is dropped. The church of the holy dollar commands the day and the mental role model comes from a glass screen. The environment is gone, passed in a hurry to nowhere.

The ordered wisdom of nature inside the global forest still stands tall. It is the majesty that beckons us to keep still and behold a beating heart in a feathered breast. The forest forecasts our future in every breath it takes and every seed it releases into the leaf mold of the forest floor. This wisdom is the universal voice of life, seeping in silence in search of our souls.

A functioning forest is a complex form of life. It is interconnected by its own flora and driven by the mammals, the amphibians and insects in it. It is kept in place by fungi, algae, lichens, bacteria, viruses, and bacteriophages. The primogenitors of the forests are trees. They communicate by carbon-coded calls and mass-market themselves by infrasound. The atmosphere links the forests into the heavens and the great oceans. The human family is both caught and held in that web of life.

The history of any particular forest in the global garden is unknown. No records exist of past splendor. But a modern mono-cropped plantation of trees is not a forest. It is a plantation, low on the scale of necessary biodiversity. Plantations are needed by farming communities for cash crop set-asides. And functioning forests are needed, too, by the global garden. They are the lungs of the planet, keeping the atmosphere rich in oxygen and low in carbon dioxide. That balance is needed for life.

By ignorance and greed the trees of the global forest are being genetically modified. This will have a domino effect that is unpre-dictable. Pesticides used in forests will kill much more than their target organism. A clean environment even within a forest is an oxymoron. Contamination is worldwide and increases with every mouthful of packaged industrial food we ingest. Toxins accumu-late and sometimes accelerate by synergism right up the food chain to the toxic top, us.

Right now there is a form of pollution in human beings, in plants, and in animals. It is an internal molecular pollution of our bodies. For instance lipids, or essentially fat in the body, are broken down by an enzymic process called beta-oxidation. Two carbons are snipped off at a time to break a fatty molecule down to zero. Many fat-soluble pesticides occur in threes, so the body is left with a single carbon at the end of this oxidation process. This polluting carbon builds up and the body does not have the enzy-mic skill to get rid of it. This contamination is in the clockwork of the metabolism itself. The answer to this is a form of reverse osmosis within the body. Clean organic food is grown and eaten to displace the contamination. There is a dilution effect. It works because all cells are mobile in their contents. Food comes in and waste products go out. This is true for trees and plants, animals and humans.

The holistic answer to pollution reversal and to stabilizing the environment is the bioplan. The act of bioplanning knits back the battered element of nature into our thinking. While it will not put the genie back into the bottle, it will contain the mischief. And time itself will form a new safety net. This is because the genetic code has its own buffer system and this system needs time to work its spell of the induction and repression of enzymes, the healing of all mechanisms of all life-forms from the bee to the boy.

To bioplan a forest, epicenter trees are chosen. These kingpin beginning species should be native to the local area. These trees must be from the highest and best genetic origin. They have to be from mother trees that are the oldest and healthiest specimens around. The genetic material of epicenter trees will hold through climatic change, pestilence, and drought. The trees will have a handful of internal tropisms or refined hormonal pathways vital to the life of a forest.

Tree seeds should be collected in the best growing year for genetic predictability. Quality seeds are superior to somatic cuttings, because seeds have an imprint to forecast future growing conditions. But increasingly, overharvesting and high-grading, or taking out the best of the best, has diminished this choice of superior genetic stock, especially for rare trees.

The epicenter trees are protected as necessary. Diversity is introduced. Ideally this should be a mix of evergreen and deciduous trees. These trees can be either seeded in or planted as young saplings. Native forest perennials can be introduced. These should be all the corms, tubers, and bulblets that would naturally be found in a local forest.

Bird boxes go up. The songbirds will introduce climbing plant species and many annuals in their droppings. Many of these woodland plant species need the scarification of stomach acids to etch the testa or outer coat of the seed. This enables germination to take place. Both birds' and mammals' annual manure provides the vital nitrogen for seeds. The other organisms, the nitrogen bacteria of the nitrogen cycle excreted by bowel movements, provide the jump-start for the essential mycorrhizal root growth.

The wind is also an important actor in a bioplanned forest. Lichens and liverworts, mosses and fern spores drift in. Sometimes gales and hurricanes distribute spores. Over time rare plants like orchids find undisturbed ground. They arrive when either the shade or sunshine is perfect for their desires. And thus a forest is born from a bioplan, slowly, succinctly, and in secrecy.

All countries and nations benefit from forests. They are places of quiet refuge. Out of every quiet thought is a rebirth of the mind in its own humanity. Trees and forests are international treasures. A forest in the global garden is a living cathedral of nature.

Climate Change Can Be Reversed:
Simplicity, Sustainability, and Sanity

SMILE, MONKEY, SMILE

The climate is currently much in fashion. It is changing. But the climate is always changing. It does it every year. It is the only thing that can really be expected, change. But this one is different, will be different, from anything we have known or perhaps even experienced as a species, this change to come.

Climate change has happened before. It was so long ago that only the rocks hold records of its effects. Strata of ancient formations hold herbaria of fossilized plants whose unique complexity is just an echo of what we see around us today. The softer rock strata occasionally spit out ossicles from ossuaries of extinct mammals, making a guessing game of what was there. All of science curtsies around the one bone fragment of flight of a prehistoric hummingbird. And all of science, too, then argues about the catastrophe that made this all come about. Today we know.

The earth, our home, is a living system. It was born. It lives. And it will die. These are the markers of life. The first marker of birth came out of the Big Bang, a theory as good as any other theory in astrophysics. It serves as a useful birth record. The next phase of planet earth involves its life. The earth speaks in a voice that is green, she moves in millennia of motion and paints the spiral pattern of design that replicates. The mirror images of chemicals and the spirals of mathematics are just one reflection of her design. These designs are only occasionally glimpsed in flashes.

The earth wears a mantle of life. This mantle is being dragged off her shoulders. It is being pulled away. This nudity will bring about the last phase, death, all deaths. And then there will be an eternity of silence.

The mantle of life, like all living systems, has a buffer. This buffer protects the equilibrium that exists in nature. It is the margin of error, built into all living systems for their ultimate protection. The atmosphere has changed. Some gases have increased. One of these is carbon dioxide. There is a rise in the concentration of this gas. The rise is unnatural. It is a toxic gas in high concentrations. It alters the life processes of the chloroplastic thermodynamic reaction of all plants. In the past, high levels of carbon dioxide and decreasing amounts of atmospheric oxygen changed the growth of fungi, mosses, and ferns. Growing within this acidic atmosphere, these species rose to heights of several hundred feet to dominate the landscape. Carbon dioxide also alters the bicarbonate buffering system of all of the oceans for marine life. High levels of atmospheric carbon dioxide induce a drought at the oceanic surface. Atmospheric moisture is needed to produce the sea bicarbonate. Shells and skeletal structures are built from this material. A rise in carbon dioxide changes the living chemistry of all soils by stealing the ubiquitous hydrogen ion, essential for all soil-to-soil and soil-to-root interaction. Its end-product effect is extinction of aerobic life-forms that need oxygen.

As the earth's temperature rises, this too changes the atmosphere, the oceans, and the land. The change is a chemical one and is common to all organic reactions on which life itself is based. The command base of nuclear DNA issues orders that must be obeyed. These orders are always biochemical in origin. These give rise to the biochemical pathways of a species' life. All biochemical reactions are governed by temperature. Lower the temperature and you slow down the reaction rate, but if you raise the temperature you increase the reaction rate. Climate change of increased temperatures will increase the rate of every chemical reaction of life. This will be as true for the pathogen as it will be for the pea.

This rise of reaction rate influences many things. The character and behavior of atmospheric gases alter. Their random pattern changes into increased chaos. This means that all atmospheric gases react at an increased rate. Which in turn increases the chance for ionization to occur. Ions can change the pathway of the sun's photons so less of them enter into the chloroplasts of the mesophyll tissue for photosynthesis to occur.

A rise in temperature of oceanic water increases the reaction rates of microscopic plants and fish and disconnects their feeding pattern both with themselves and with the sun. This wreaks ecological havoc in the oceans. A similar prognosis is valid for soils where the beneficial microorganisms are put into imbalance.

An increase of temperature affects all vectors of disease of plant and animal origin whose ability to ravage is dependent on the ubiquitous chemical reactions of both the disease vector and its prey. This is valid for equatorial diseases right into both poles. The temperature increase will help disease to move and shift into new neighborhoods on the globe. It will also help many diseases to initiate host-hopping into new domains.

The vast areas of tundra, too, will change. These cool regions sit on enormous quantities of bonded carbon dioxide. This carbon is in an organic form of decaying plant material in both dry and water-soaked areas. An increase in temperature increases evaporation and the decay rate of these regions, thus releasing ever increasing amounts of carbon dioxide into the atmosphere.

There is an answer to stalling and reversing climate change. It involves the use of the triple-S system. This calls for simplicity, sustainability, and sanity: simple living twinned with sustainable thinking and fine-tuned to the wholesome sanity that we can all do something about this problem. If we cause it, we can also cure it. Smile, monkey, smile.

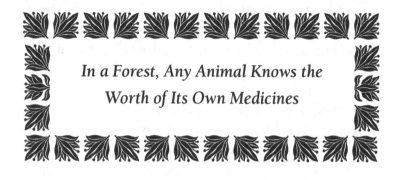

In a Forest, Any Animal Knows the Worth of Its Own Medicines

THE MEDICINE FOR MAMMALS

Little is known about the secret life of mammals or of birds, insects, and fishes either for that matter. But every species has its own medicine; from the bacteria the fungus-farming ants carry for the production of antibiotics to the licking machine of fur-bearing animals to increase circulatory flow for soreness or damage.

The great forests of the global garden carry many medicinal plants for wildlife too. The bounty of medicines is not lost on the creatures of the forest that seem to have an imprint of this knowledge much like a hen laying her first egg. It comes as a surprise, this slippery white oval thing, but it is hers with a full claim of identity although the shape in no way resembles the hen that laid it.

There are extraordinary stories of medicines of the global garden and their relationship to plants and animals. One special case is a native North American daisy. It lives in the center of the continent in the Great Plains area. The daisy is called *Echinacea purpurea* because of its prickly similarity to hedgehogs and their cousins, the porcupines. This plant is also known as coneflower. It shares a dry habitat with eight species of rattlesnake. The coneflower is an ancient medicinal plant with rather extraordinary biochemical antivenin power for each of the eight local snakes. The biochemical comes from the plant with a caffeine molecule attached, so that entry into the digestive system across the stomach wall and into the circulation system is rapid. This rapid-fire is necessary to block the poisoning effect of the snake's venom and save a life in the process.

Canines—wolves, coyotes, and the family dog—quite often forage for their own medicine in troubling times. The dog searches out a species of grass by smell. This grass grows alongside other cousins of the *Gramineae* family. But the dog can pick out the special one needed. It is *Agropyron repens* or couch grass. This grass, in its mature state, contains antibiotic mucilage mixed with triticin sugar. Ingested, the small pieces of the blade of this grass will clean out the urogenital system, which is much in use by these creatures for territorial marksmanship. Such tidings are the mark of good housekeeping in the print of a fresh pee.

In the tropics even the hippo has a surprise. The hippo heads for tyrosine-rich plants and stores them up. They later get converted into two pigmented acids to produce the hippo's colorful sweat. The red-orange sweat acts as a sunscreen against UV rays and it is also an antibiotic for the hippo's hide. The odd tangle with another hippo could produce grave results for a flesh wound in a hot and humid environment. But the antiseptic effect of the two skin secreted acids looks after that eventuality.

Back home in the northeastern American forests another smorgasbord of biochemicals is on display in the fall. One of these is a hemostatic agent called juglone. This medicine rapidly stems the flow of fresh blood and acts as a clotting agent. The entire worldwide *Juglandaceae* family of walnut trees carries this chemical just inside its trunk in the cambial tissue. The deer seek out these species of trees. It is a favorite of the buck when his antlers begin to itch and the velvet covering becomes loose. The buck will rake his itch until it bleeds against the *Juglans* or walnut tree species, until the cambium is exposed and the juglone is available as a clotting agent. Juglone also acts as a sterile swab for the newly exposed pedicel on the top of the skull after the antlers have abscised.

Probably some of the most important biochemical medicines produced by the global forest go unnoticed. These are the waxes, gums, and resins of the forest. These are bled by the trunk of the tree in resin canals and sometimes they are found near the elongated, lens-shaped breathing pores, or lenticels, of the trunk. They are also extruded by special glandular tissue of leaves and of petioles. Sometimes these compounds are colorless and other times they are colored. Many are scented and have been used all over the world as chewing gums.

The waxes, gums, and resins are polymers or they are mixtures of polymers bonded with chemicals that shed water and are hydrophobic. This wealth of chemistry is remodeled by heat into the propolis, the glue, of insect homes. This propolis keeps the homes both watertight and sometimes airtight. They are the defense chemicals of the insect world. They also have strong fungicidal and antibiotic action. They are the medicinal wallpapering of the insects' homes.

On a hot and humid summer's day a worker honeybee will go in search of resin. She tears away at it with her mandibles and brings it back to the hive. She stops her flight at the landing board and there tiny splats of resin can be seen. The resin is carefully mixed into the waxes of the hive body and is used for sealing the hive against invasion and for controlling the temperature inside the hive. The resin is also used to encase another insect that happens to have entered the hive. It is entombed in propolis to keep the rest of the hive sanitized.

The life of an insect pollinator may seem to be very tiny and insignificant indeed, but these creatures multiplied into the billions are a key part of the link in the chain of life. Without their health there would be no pollination. Without pollen the seeds would not form. And inside each and every tiny seed of a good food crop is a kitchen connection to the family and the promise of another spring to the farm.

Trees Make
Exemplary Mothers

OH MOTHER, DEAR

The mothering instinct comes naturally to a tree. Trees are like warm-blooded mammals that will protect their offspring, in some instances to the death. But the tree does it differently; the fight for life is with the weapons from the mother's ancestral arsenal.

Like most mothers, most trees are not naturally solitary. They are community dwellers. The community for the tree is the forest. Inside the forest all mother trees get the greatest protection possible. It is in the form of sophisticated alarm signals. These are generated by rapidly moving carbon-bearing molecules that flash out from the tree spreading into the atmosphere. They move in airways around the trees of the forest. This molecular movement alerts the beneficial bugs to move in to help with predation. Trees also swap carbon from one to another underground. If one tree is not doing so well in the shade and carbon is in short supply, a bigger nearby tree steps in to help. Carbon swapping for feeding is carried on by soil mycorrhiza or fungi, whose hyphae stretch immense elastic distances in the forest floor.

A mother tree, when she is in seasonal gestation, has her wishes and wants. Many of these are manufactured by the tree itself. Some are initiated as fragrances, when pollination is about to take place. The gynoecium, or female, ova-bearing part of the flower issues an open invitation to pollinate by injecting fragrance into the air. These chemical messages are corralled by beneficial insects that move in for the buzz. The pollen is liberated and spun free at this time. The ripe pollen, too, holds its own identity message for cross-pollination of like species. And the mother tree gets what she wants, fertilization.

Occasionally a touch of cannibalism takes place with some trees. If the nitrogen stocks are low on the ground, the blue-green algae component of the mycorrhizal membership becomes severely stressed for it. By the light of the moon the blue-green algae switch on their nitrogenase stable of enzymes for available nitrogen to make protein. If the nitrogen is not present or scarce in the ground, the mother tree moves for her arsenal. She increases the concentration of her fragrance and puts a hypnotic chemical into it. The beneficials, like bees, become comatose and die. Their bodies add the water-soluble nitrogen much needed by the tree for sexual fulfillment and progeny development.

The progeny or the children of trees are wrapped in swaddling clothes. Usually there are two layers for seed protection, an outer and an inner. There may be more, too, depending on the tree species. The usual outer layer is hard and tough to break. This normally gets the seed through the hurly-burly of winter. This coat is called the testa. Then there is a finely designed inner vestment that clings to the meat of the seed. This intima senses dormancy and future division.

A good mother tree that has been around for a long time also sees that her children can defend themselves. She provides them with their own medicinal kit. Occasionally the seedpod in which the seeds travel to their new destination has antibiotics or fungicides for the young seeds' protection. Other times there are antifeeding compounds, or maybe for some species of trees there are enticingly delectable chemicals of intense sweetness inside of the pods. These pods are destined to travel through an intestine somewhere.

Sometimes the seeds have a trigger for smoke, and others have a trip mechanism for thunderstorms with their quick flush of available nitrogen fertilizer. There are seeds with downright dangers like neurotoxins and there are seeds with delights for either man or beast.

Sometimes the mother tree takes care of rivals in her own patch. She does this by secreting a series of complex allelochemicals into the soil around her skirts. These chemicals can move in the soil and become deadly to other saplings or any other species the tree may take a fancy to kill off. Many trees of the pea family use a form of "nuclear" warfare for the seed's defense. All invading pathogenic fungi that are a threat to the pea seed's happiness receive a nasty notch cut into their DNA. Occasionally the allelochemicals are airborne from fallen leaves as air sterilizers to counter infections.

Trees and plants have never been thought of as warm-blooded like mammals. But they are in fact. As the spring sun rises in the southern horizon the trees warm up, beginning at the trunk. There is a black box passive effect but there is also a clear increase in the metabolism of some trees over and above this passive status. This warming phenomenon is used by the wildflower populations of tubers, rhizomes, and cormous plants that hug the apron area of the trunk. This early awakening initiates a release from plant dormancy. In many cases this is aided by insects undoing the sugar glues of winter that tie up buds and flowers, so growth can commence. Big mother trees do this careful spring warming better than smaller trees.

In a forest the best mother trees are the healthiest and the more mature. Often they are the largest also. These are the trees that have learned the tricks of the trade in living. These trees carry the best card for genetic deliverance in an adaptable light- and climate-controlled enzyme system. These trees also carry the best medicinal tools, triggered to last centuries, if not millennia. Mother trees of the highest caliber produce offspring of a similar rating. Despite all of this, it is the mother trees that get the axe in the global garden. They do not receive the respect that they deserve, ever.

The Forests of North America
Wear an Apron of Flowers

THE FLOWERS OF THE FOREST

The luck of the draw for the Americas came with the aboriginal people. It was a good day for North America when they landed forty thousand years ago. Their reverence for the natural world around them saved so many of the precious flora. The abundance of their reverence is in evidence from early spring into fall in the countryside and in the forests. This is not so in Europe.

The treasure trove of medicinal drugs used by the medicine men and women of North America was a spectacular one. Medicines of the margins of nature are always superior because plants that grow under stress produce a large alkaloidal arsenal. The alkaloids in plants are part of the universal fright and flight strategy for survival. The biochemistry of these alkaloids is unique. Somehow through keen observation this became known, and the harvesting of plants was always with a view to the plant's having been around to a minimum of the seventh generation, an oral formula for survival of all species.

From early spring the forests of North America wear an apron of flowers. The first flowers in every forest are served by the trees, and they have much in common with one another from forest to forest north and south. The early blaze of bloom arises from underground roots. Sometimes the flower emerges without the benefit of leaves as the early bird in the annual race of life. The roots are always modified, carrying food reserves from the previous year. These roots can form corms, tubers, rhizomes, succulent areas, or enlarged flower platforms. The previous year's precipitation gives rise to the flower's size; a drought will form a small flower while sufficient rain makes the flower lush.

The early spring flowers are always found hugging the trees of the forest. They gain a jump-start in growing from the warm-blooded behavior of the tree's trunk near the ground. This additional warmth from the tree heats up the soil. In the more northern regions the trees actually melt the snow and ice and shield the early flowers with their black box effect of solar radiation. Many of these early flowers are heat generators, trilliums and hepaticas in North America, fritillaria and hellebores in Europe and Asia. This heating fans out the bloom and mobilizes fragrances, which initiate early pollination from the first flying insects. These pollinators can sometimes be seen clinging to petals waiting for an upturn of temperatures. The actual flowers are always fleeting. This is strictly energy conservation for them. They always leave behind the roots filled with food.

The next tier of flowers that arise in the forest are always more brave. They will be found farther away from the protection of roots and with this have other unusual characteristics. They will have a thick and sometimes shiny cuticle on their leaf surfaces.

This ingenuity is for water control. Out in the open the flower is more prone to evaporation and water loss. A waxy surface on the leaf helps the plant to conserve internal water, and a shiny surface helps the drip drop into and around the root system. This cleverness in drainage puts power into the flower. The blooming period of these independent flowers such as wintergreen, wood poppies, and wood mint is always a little longer, as if these swaggarts knew their own good fortune.

Summer moves through the forest with a nearly empty basket. That is because the trees themselves are echoing the toll of fragrance, fertility, and pollination. All life moves up the canopy in an answer to this call. The flowers that are in the forest in the heat are the mobile ones. These can flaunt their wares in one corner one year and then move to another site that is better situated for business. These are the annuals with room to spare in their lives. The somber perennials are almost finished now. The annuals with their gay abandon of seeds from the previous year needed a head start for germination. The young seedlings lift their cotyledons amidst the crinkle of leaves on the forest floor using the heat and protection these leaves can give. Then when conditions are correct the annuals will rise in glory to bloom. They will cause a downward flow of traffic with insects and butterflies.

Late summer and fall is seen in the forest by the next succession of another force of nature. The ferns lie resplendent in their sweeping thrones of buttressed roots. And all around them arises a stippling of impressionism, clots of mushrooms fresh with vivid color. The ferns are flooded with fronds. Under these fronds tiny pockets of sori are filled with soft brown spores. These patiently wait for the dry winds around the forest floor. And one morning as a drop of dew lands and leaks its vapor a sorus will explode with spores, catapulting them into their new cradles of soil and rebirth.

All the while the still stippling increases in color. With glaucous maturity the young caps of mushrooms break open the soil heaved up by the hoards of underground hyphae. Each species of mushroom has an intimate connection with its own species of tree. They have lived together underneath the soil for a long, long time. And now the fertility rite must take place. The reasons are unknown, a mystery of nature why this sexual union of a subterranean marriage will arise and be seen as a mushroom fruit. Each mushroom, too, will play a waiting game of suspension. And one dampish or maybe even one extraordinarily warm and windy day a puff of smoke will be seen. It is the inky chain of spores, rising into the air as a cloud to wind its way around another passion in the privacy of the forest.

The vascular war of the worlds is on. The vascular plants represent only one-tenth of what was there before. The current number of a quarter of a million species is getting smaller by the hour. The majesty of diversity is being reduced by ignorance, and nature is being assaulted in all corners of the world, and now, it would appear, in the boreal forests, too. The wars of the past were wars of nations. Now it is corporations that slink into the slot. And the flowers of the forests are being trampled . . . every single day.

*Art Is the Telltale
of the Forest*

OF FORESTS AND ART

Art is the compulsive search for truth. It divides man from the breast of the beast in the great world of mammals, for art defines the conscious man. And art throws those thoughts of man into the exploration of truth, into all the possible forms through which art can be illuminated to others. For art is the mirror of the mind in the journey into that truth.

Of the graces of mankind art is one. It is present at conception in knowledge. It flows from that knowledge into the young child in a stream of activity. That stream builds recognition of self and its product is unity. A knowledge of self forges the man into mankind, while it fosters his unique separation, too. Such things are spoken of as the "maturation process" and are the basics of the ability to live in peace beside others in concepts of civil society.

Art is the great shadow of man. It has dogged him for a long time and is seen in many places. It is seen, too, in surprising things like tools, weapons of war, and clothing. Ancient art is to be found in cave paintings, in petroglyphs, in weaving, in woodcarvings, and in intricate designs. The face of art floods the races of the earth from the beginnings of time.

But art also has a sister. The sibling is science. Art and science are of the same house, of the same family. Art in all of its forms opens the way for science, because art is a precursor to science in all things. Art sounds the bell of change that leads to discovery, and science runs in to listen, to test, and to learn. Art sometimes molds and other times reflects the thoughts of culture and then defines the tides of fashion. Science follows in the wake of those tides and looks back at that great fetch of "why" to derive the question "how."

Creativity is the wellspring of art and science. Nature takes a share in this too, because creativity is the foundation force of nature that drives all of the processes of life. It also drives the inter-connectivity within the species of the spectrum of life. This is what regenerates nature and gives it fresh and renewed life. Art, too, is defined by its creativity and would not exist without it. Science depends on the inspiration of creativity for its thought.

The creativity of nature is a constant. This value is found in the almost eternal flexibility of nature and its ability to massage all life-forms that exist. The creativity of science comes from an enlightened plagiarism and observation of nature. The creativity of art tracks nature itself and all of the abstractions bled from it.

Art and science are also about the discovery of truth, the universal context of nature. If truth emerges in art, then that work presents a form of unity of majesty, which can transcend the art itself. This is seen and addressed by culture over fashion. But if truth is found in science, then an understanding of nature and of its spectrum of species is advanced. For truth is the tool of science in discovery.

A truth is emerging from art; it is walking out of modern works with a spectral presence. This holds a message for the peoples of the world. That message itself is a signpost toward our own human destiny. The message is that nature itself is being degraded. It is being picked apart. Transmuted. In the art of the last handful of centuries the global garden was seen to be massed with mighty trees and landscapes of roaring rivers and pristine mountain peaks.

The art of these past few decades shows saplings leaning in harmony over areas of past spoilage. The forest with its exotic presence has gone. The vision of the virgin forest has gone, too. The cathedral of the forest, ringed with diversity and splashed with light, will soon be gone with the amphibians, the great mammals, and the fish.

The civilized world has not put a finger on the pulse of nature. It has ignored the pattern in which nature works, as if man himself is an independent species apart from the web of it. The truth is that man is only one species and he stands on a fragile platform of life that is but a whisper away from death.

There is some time left. There is time for a different way of thinking in which man can rethread the needle and sew a life for the future. For if nature is destroyed, art will stand still and the creativity of science will follow suit. Civilizations have risen and fallen many times before. This time around, it is different. Now we have the lessons of history behind us. Let us look into ourselves for a new face in art . . . and another one in nature.

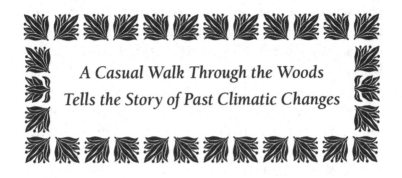

A Casual Walk Through the Woods
Tells the Story of Past Climatic Changes

A CHAT ABOUT CLIMATE CHANGE

There has been much observation of the atmospheric pulse of late and all of it looks grim. The patient has a fever and the clinic is on fire. The doc's gone. He won't be back.

Before the advent of trees there was a climate, too. It had the usual cycles of ups and downs that swept by, counting the millennia. The populations were prokaryotic. Some had a mind toward advancement but the weather got in the way. The carbonic rain was a problem and so too was the lack of oxygen. First the mosses and ferns came along. They dragged their heels until the gymnosperm took over. Then the trees looked in on the scene of population explosions and decided to hang around and keep sucking up carbon dioxide out of the air. They added back the spit and polish of oxygen. Man and his babies stepped into the picture. They were a gift from God, so they thought, as they arrived.

Our global village, our home, planet earth, has all of the astonishing aspects of a living unit much like a cell. It has its own respiration curve based on the seasons. It has its own investment capital called carbon. These capital reserves are shared with the oceans and atmosphere. They are in flux. As of the last 200 million years much of the carbon reserves have been vaulted in live chloroplasts and also in fossil form. But these are also traded underground in a lively exchange. Sometimes there is a bear market for carbon and other times a bull. All of the excess carbon is buffered in treasury bills, here and there. The banking system of planet earth is old and has been a successful one. It has been based on prudence, the first commandment of bankers. But above all it has been grounded as a sustainable system on which, after all, quite a lot depends.

Lately the oceans are getting warmer and expanding. The polar ice caps are melting. This story is that it is just the beginning. Glaciers are breaking up, opening small seas of ancient freshwater lakes trapped behind them. Beaches, rivers, islands, and countries are being drowned. But wait, the weather is involved here too. The sun and the earth have been involved with each other for a long time. It is a seasonal thing. The sun moves up out of the southern horizon into the north and we call it spring. In the oceans this causes the feeding plankton to snap into life. They begin the serious job of photosynthesis. Then they divide. They proceed to explode in numbers. These are all the busy bottom feeders. They are the diatoms and desmids, the unicells and the multicells. They are the cyanophyta that show sexual expression in their ribbons. They all respond the same way to the sun, their populations explode. They are the bottom of the food chain.

There is always somebody ahead in this pecking order. These are the tiny crustaceans. These dine out in massive numbers and get so numerous that the fish larvae take notice. They too begin a feeding frenzy. The fish grow fat. They are eaten by bigger fish, and so on. This is life as a fish knows it.

In the last fifty years this has all changed. The sun comes up on the horizon to announce spring. The nanoplankton receive their charge of photonic energy to free up the mobile chloroplasts. But the oceans have become superheated. The temperature of oceanic water has increased by 0.9°C, almost a full degree centigrade. This has changed life for the crustaceans because these creatures are temperature dependent and want to be fed on an earlier schedule because spring has come to them in the form of warmer waters. They are ten days too early looking for food. But they are out of luck because the nanoplankton are only solar dependent and nothing will push them into a division for food mass over and above the sun. The crustaceans starve and so do the little fish, and this concertina expands up into the bigger fish. And this is called a collapse of the oceans.

Then another thing happened while there was a rainy day. The fossil fuel reserves of carbon were slated for the car industry. The cars burned all the fuel up as combustible material and released the carbon dioxide back into the atmosphere. It stayed there because all the forests were being cut down and so the factory for carbon dioxide sequestration was dismantled. Nobody is seriously stopping this. The carbon vaults are being plundered, and this too is putting even more carbon dioxide into the atmosphere. This is called a collapse of the forest system.

But there is some good news on the horizon. People are talking about climate change. This is happening everywhere. The conversations are in languages too numerous to count but the idea runs as a common theme. "Climate change is happening and what can we do about it?"

This is the generation of the greening of the globe. We are at the end of the old ways and the beginning of the new. People are ahead of the politicians this time. Ordinary people in simple homes eating bread and butter are about to turn the human herd toward a new destination. They are doing it because the herd is in danger. It is a collective instinct as ancient as life itself.

This time there is a democracy of thought. Pollution affects everyone. Nobody escapes. The soil must heal. The atmosphere has to improve. Smog must go. The oceans have to be protected. All water needs to be pure. Clean energy sources can and must be found. Biodiversity has to be maintained. Forests will be planted. More forests must be planted. And the trees will smile their oxygen again in the dawn of a brand-new day.

A Need for Pulp
Peels the Forest Apart

THE PAPER TRAIL

A comic opera is in full swing. It is happening today on the stage of our lives. The blind man in costume has climbed the tree. He carried the saw in between his teeth. He is sitting on a branch. He is in full view. The whole world sees him. The branch is shaking. He has moved his backside to get a better purchase on the handle of the saw. He is cutting. He is sawing away on the limb that sustains him. Maybe in this comic opera he is meant to fly. And then maybe he himself expects to fly. . . .

As this society of ours edges "down the primrose path to the everlasting bonfire," we say that we are becoming paperless. Nothing is farther from the truth. Every spoonful of sugar is wrapped in paper. Every meal and every drink. Every move we make outside of the home and inside of the workplace involves an assault of paper. Landfill sites are overflowing. These are costly to the taxpayer who is now expected to pay for incineration whose airborne particulate test is opacity. And so the atmosphere is filling, too, with nanogram amounts of fibers from paper products. Trains and gigantic trucks are transporting new and used paper from pillar to post, in and out of cities, from nation to nation in cash-sweet deals. The demand for paper is exploding to over 200 million tons of pulp per annum for the Western world. We need not worry; all is safely in the hands of the corporate world.

Pulp and paper come from trees. And trees come from the forest. Trees make up the global forest. Not every tree within the forest is a good pulp tree, though all trees can make pulp. The pulp itself that is used for paper is composed of fibers. These fibers are long xylem cells that make up the internal circulation of a tree much like the veins and arteries of the circulation system of mammals. A fiber is either an early wood or a late wood tracheid; these fit together to make the vertical plumbing of a tree. They can be seen in a cross-section in more porous wood as tiny little holes. They can be counted en masse as the tree's annular rings. Every interested child and every captivated biologist makes this counting, a game of sight.

Trees are shorn off the landscape by six-million-dollar lawn-mowers. They are mauled and hauled by the feller-buncher until each tree is a battered trunk. They are collected and retted until they are digested down to their fibers. And the fibers bulk to pulp. This pulp is realigned by dehydration into the smoothness of paper. With all of its variety of quality and fineness it strolls in to meet industrial demands of a civilized and literate society.

Trees have been the most plentiful, the easiest, and the cheapest source for paper. The forests of the global garden have been ripped down as if the trees were self-spawning in some vast self-sustaining cycle of growth. In the 1950s the global surface was 30 percent covered with forests and in 2005, it was in the range of 5 percent.

The great northern boreal forests are now being cut. And our tomorrow will drag in the image of Easter Island for a final viewing.

It is not necessary to use trees for pulp. All vascular plants can produce pulp. And some can do it better than others. There are 250,000 plants that can make pulp for paper. Unfortunately this just represents 10 percent of what was there before the various extinctions of past climatic changes. Some of these vascular plants are outstanding for fiber quality and others can be processed by heat to strengthen the fibers for a further recycling capacity. Most of these vascular species have been used in the past in North America and in Europe. One of these is the urban criminal *Cannabis sativa*, hemp.

It was just a few years ago that an insipid attempt was made to exploit the urban criminal for the very fine fiber this species produces in poor soils even with short northern growing conditions. This repast of crime happened in eastern North America in a continent ideal for growing this scamp because of the ample solar conditions to which this continental face is predisposed.

The farmers obtained the seed. Fields were duly tilled and the crop was planted. All of the crops everywhere began to show the excellent characteristics of growth typical for hemp on poor soils. The plants reached for the sky. They grew bigger and better than any crop any farmer had seen in his or her days. The fields were visited in a patter of pilgrimage by people who saw them as wellsprings of hope for the future of their farms and the environment for their children.

The *Cannabis sativa*, hemp, appeared to be top notch for pulp. The government held its collective breath. The days drew close and the combine harvesters drove into the fields and began the task of cutting. But the machines, all of them everywhere, got clogged with resin. Resin is not cleaned from machines with a garden hose. It requires a suitable solvent, one that can easily be engineered into the machinery's design. But farmers are not engineers, nearly, but not quite. The farmers were left holding the bag. Some went bankrupt. Others went to the brink. And the government, to a man, went silent.

And so the comic opera runs to an end. The branch is about to give. The blind man in costume looks up, not down, because he has heard the breaking of wood underneath his backside. He looks up to the sky quickly again and wonders . . . "Is it too late to learn to fly?"

The Global Forests Erect Chemical Atmospheric Barriers Against Bacteria, Pathogenic Fungi, and, Possibly, Viruses

FRAGRANCE

In the salons of nature, fragrance is the calling card. For the forest it is the chemical messenger that can travel miles for delivery. Often the messenger has the benefit of airborne aid by other high-flying molecules. Once received, the message of fragrance can initiate many things from health to death, identity to recognition, misery, avoidance, and even sterilization. Fragrance can come in many forms, from ethereal perfume, neutral odors, traces of scent, and pungent fumes to carnal smells of rotting flesh and other mixtures of unbelievable stench.

Trees and many plant species that produce fragrances have evolved a strange kind of cunning. Within the fragrance there is a mixture of chemicals released. Many of these have a benefit for a target species. These are medicinal fragrances and have great value and even greater interest. One of these is the blue *Monarda didyma*, bee balm or Oswego tea. This is a native North American species of the virgin forest. This plant, like many species of trees, produces bergamot, an oil which is dispensed with a carrying aerosol.

This delightful fragrance opens out the lungs as a bronchodilator when the pollen levels are seasonally high in the forest. This sweet-smelling native keeps the lungs clean and healthy. It is a medicine.

A fragrance, an odor, a scent, a stink, or a fume all function in the same way. The chemical that is released becomes airborne like a balloon. This can be light or heavy. If it is light it will climb into the winds and fly. If it is heavy, it will not go far. These chemicals are unique because as they are released they change their character or their chemical identity and leave a little of themselves behind. The bit that flies is called the aerosol; it moves and dispenses rapidly. This aerosol holds the fragrance or the smell that becomes the calling card and spells out the actual message for the receiving species. These are almost always important medicines.

A chemical of fragrance is an organic chemical. It is released by the tree as part of a biochemical reaction. These reactions are temperature dependent. In winter they slow down to almost zero. As the temperatures gear up for spring and summer, these reactions increase in rate and number. This is the smell of spring.

Trees, because they are tall, are highly effective chemical communicators. A tree's height gives it a greater ability to disperse or inject chemicals into the airways and passages around the forest and the atmosphere above it. The aerosols from a forest are emitted from the chemistry of the canopy itself, sometimes from the flowers, other times from the extra floral nectaries or resin canals. Quite often the fragrances are held in special organs called glands or glandular tissue. These glands are microscopic. They are a world unto themselves. They are found on leaves and occasionally on the petioles attached to them. They are found on young branch tips and also on the husks of nuts and fruit and occasionally in the fruit itself as flavor.

Scent glands more commonly are composed of fine hairs called trichoma. These are always found on surfaces and make a tree leaf look like a miniature jungle. These hairs function like booby traps; if the tips get injured, the fragrance is released. Other glands for fragrance are rounded. These are rigged with an internal pressure system and are explosive like a land mine—the skin of an orange is fraught with such land mines. The fragrance and its spray from an orange skin contain orange-smelling, phytotoxic agents that prevent any fur-bearing animal from pilfering or peeling it. Most birds will stay away from an orange for the same reason.

The more familiar fragrance of the forest is that of the pine, *Pinus* species. Hence the worldwide use of these trees in health spas. On a sweltering summer afternoon the pine will emit an odor that is detected at ground level. This odor is a medicinal mixture of various esters of pinosylvin. These and other aerosols are from the truly ancient pharmacopoeia of the pine. The pinosylvin is a natural antibiotic. When emitted as an ester form it exerts in the forest a stimulating effect on the process of breathing itself. It also functions as a mild narcotic. These aerosols have an anesthetic effect on the body, bringing about relaxation. A forest of pines acts as an air sweep, cleansing and soporifying the atmosphere anywhere they grow in the global garden. Other trees do likewise. In fact the global forests exert an antiviral and antibacterial action on moving air masses, in general.

There are other trees that do the opposite. They operate a nice little sideline on death. These are of the poplar, *Populus* species. If a surface root gets nicked and the bark is torn off, this sets up an aerosol alarm. The higher order of fungi, *Ascomycetes*, come in to help. Some of these fungi have a sexual organ, which is identical to the male penis in full-blown erection. These are called the stinkhorn or *Phallus* mushrooms. From the glory of these stinkhorns arises a raunchy odor that goes beyond description. It crushes the brain. These fumes linger, motoring around the pink phallus especially in the morning hours.

These penetrating fumes advertise a chemical message of death and decay. The smell ripens to that of rotting flesh. Soon a series of undertaker beetles arrive on the scene. They are somberly dressed in black, their exoskeletal works polished. They get down to the business of eating the phallus corpse. They clean up, licking the exposed white flesh of the tree quite clean in the process. The sober surgery helps repair the tree. The cortege moves on, to possibly better and fouler forest smells. It is all just a day's work. And some work is definitely better than others.

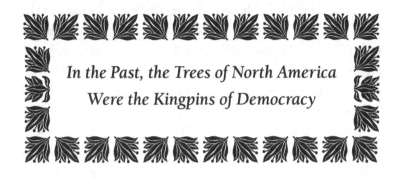

In the Past, the Trees of North America Were the Kingpins of Democracy

TWO-TIER AGRICULTURE

Two-tier agriculture is as old as the hills. In the global garden it has many faces.

In Ireland it involved a form of permaculture in which a hay field was used in perpetuity. The hayfield grew a mixture of perennials and native grasses, whose protein ratio was correct for a July cut. Then a second crop was harvested from the same field. This crop grew around the perimeter. It was a nut crop of hazelnuts, *Corylus avellana*. These large green nuts were collected from the hay wagons. A third crop was then collected from the same field with the coloring of the willows. Whips were cut, dried, and treated. These were woven into baskets used for all aspects of the laying flock's comfort including egg storage for market.

In medieval times in England, coppicing was the order of the day. The forests were cropped and kept shorn in a seven-year rotating cycle that provided charcoal and wattle for wattle-and-daub construction. This is an old building method that used clay plastering over woven whips or sticks on timber frames. In addition the forests provided oak mast for foraging swine herds and the many woodland perennial herbaceous species that flower in the wax and wane of a seven-year cycle. The native wood strawberry, *Fragaria vesca*, also unfortunately a sweet choice of the highwayman, was an expensive item in Europe's kitchens of the rich and famous. It was a valuable crop.

In North America, in the past, the aboriginal people practiced two-tier agriculture. One form, of course, was the masterful savannah, which has never been surpassed for the simplicity of its genius. In this vast parkland of edible nuts and canopies of lacelike loveliness, where protein from animal was matched to protein from vegetable to feed the continent, there existed a unique tribal understanding of ownership. There was a decided democracy of trees where each producing nut tree was owned apart and separately from the land on which it grew. The result of this was an intermingling of values whose common denominator was stewardship under a divine covenant. This underlying fact was true despite the raids that took place within the First Nations peoples. Stewardship still remains an anchor for the aboriginal peoples of North America of today.

In the warmer regions of the world where sun and heat were combined with a source of fresh water, a two-tier agriculture arose that was truly extraordinary. Knowledge of this was extensive in the global garden, stretching from ancient Japan to Lake Chad in Africa into downtown Mexico City. From the stretches of water, fresh fish were netted and sold. Then, as the summer peaked with temperature so did a minute blue-green algae called *Spirulina*. These tiny ribbons of blue-green cells multiplied into a formidable mass that was rich in nitrogen. This pungent mixture was netted in summer and dried down in the sun. It was then curdled with some form of native plant rennet enzyme, probably from the *Malvaceae*, commonly known as the mallow family. Then a delicious bread was baked whose flavor and taste was that of a most fragrant cheese.

The question of two-tier agriculture also arises with the onset of global warming and a need for clean, first-class, healthy protein. The savannah design solves many problems in food production. This design can easily be adapted to any working farmstead in North America. A large variety of native trees can be considered. These feeding trees can be grown alone or in any complementary fashion to accelerate and double the protein-producing capacity of the land. The net revenue to the farmer is self-evident.

On poor, dry land the nitrogen-fixing tree of the pea or *Leguminosae* family can be grown. One of these is *Gleditsia triacanthos*, the honey locust. There are many sugar-rich cultivars of this tree. The crop from it fed the northeastern bison of the past, ensuring that the cows had sufficient milk for the newborn of the herd. The pods can be harvested for sheep and dairy ration, and the dehydrated seeds, the size of a broad bean, produce a protein-rich and vitamin-filled flour for milling. The flour blends well with other flours in baking.

The walnut family rivals corn for growing ability. *Juglans nigra*, black walnut, *J. cinerea*, butternut, and all of their cultivars produce nutmeats for the vegetarian and commercial markets, as do the hickories, *Carya* species, especially *Carya ovata*, shagbark, *C. laciniosa*, kingnut, and *C. illinoensis*, pecan. These hickories have excellent local cultivars and cross-bred hybrids, some of which bear very large nuts, for which there is an ever increasing market and demand. Many white oaks, and many native North American western pines, have excellent nuts. These nutmeats are all rich in essential fats. These, too, add pennies into the farmer's purse.

Two-tier agriculture does the farmer another favor. Placing trees near a field with any food crop ensures the presence of feral native bees for the cross-pollination of that food crop. Bees need the diversity of pollen and waxes that trees supply. Many field crops require an injection of mixed pollen for hybrid vigor of the seed. This, in turn, results in larger seed size and increased harvest. It is estimated by scientists that these feral pollinators will increase an average farm crop yield by about 20 percent.

In addition, two-tier agriculture acts as a brake for water runoff. Transpiration and evaporation from the mouthparts or stomata of working trees feed an upward flow of water in the porous soil that aids in vertical water movement. Such groundwater, held in place by electrostatic forces of hydrogen-to-oxygen attraction that are called the van der Waals forces, stabilizes the aquifer into working columns of water. These columns of water are then less inclined to flow in runoff and, with their load of nutrients, form dead or hypoxic regions in our oceans that are growing in number at a frightening speed. Oceanic areas that do not carry sufficient oxygen to support life are hypoxic. All fish and marine mammals in these areas die. These are the new graveyards in our living seas.

The Greatest Sin for Future Generations Is the Reduction of the Biodiversity of Trees

"LET THERE BE LIGHT . . ."

There is a simplicity to the sun. It is a golden disk. It sits in the sky for every eye to see. The sun produces light energy. This energy of light is received and used by people and by plants everywhere, or almost everywhere on this planet. The sun, in a fundamental way, rules our lives from one day-length into another in our necklace of time.

There appears to be simplicity in sunlight itself. It can be seen coming as perfectly straight lines, shining into a dusty room and holding motes of dust in the stream of light. These lines of light illuminate the world; give perspective to landscape and even a meaning to its absence in shadow. This light shines on the bodies of plants and animals and it is absorbed by them in specialized organs called photoreceptors.

Photoreceptor organs occur in all plants except, perhaps, the fungi. They also occur in mammals and man as the pigmented retina of the eye. Mammalian photoreceptors scan a shorter length of the light spectrum and for the most part are more effective in light capture. Insects such as butterflies, bees, and wasps register invisible color. Many go one step further and can read polarizing light. Plants with their green-colored, spongy mesophyll and extraordinary pigmentation read the full spectrum of light, including the regions invisible to man.

In the world of plants, photoreceptors are, for the most part, the green of the leaves. They are also the range of colors that emerge with the shortening days of autumn. This, too, is seen echoed in the oceans. The green of the marine plant world on or near the surface gets the full splash of the sun. As the plants recede deeper into the ocean depths to live out their lives the colors of autumn become necessary for amplification of that much-needed light, so green species are followed by brown species of algae that can endure some depth; then by yellows and finally by the reds who are deeper still and who are so clever at trapping and amplifying the sun in their dark watery homes.

The eye acts as one set of photoreceptors for mammal and man. These are the only truly colorful parts of the body in their brown, green, and blue. Eyes are arranged to receive sunlight and to transfer the meanings of the messages they record to the brain.

Some creatures like cats have extraordinarily adaptive eyes for light. There are other photoreceptive systems in mammals and man that are enzymically based. These are fired into action by trigger flares of individual light spectra like flashlights. These are some of the most important brain regulators in man.

These systems also occur in plants and have a not too dissimilar function of regulation. But in both plants and animals, these enzyme receptor systems are coregulated by the sun, in the circadian cycle of the night and day. They are found too in the seasonal drive from short days into longer day lengths north and south of the tropics.

Sunlight travels in pulses too. These pulses are sine waves like the incoming waves of the sea, which also have both a linear movement and a pulse.

Sunlight in this pattern of movement charges the waiting voltaic system of plants. This voltaic system is made up of chains upon chains of aromatic polymers. Each polymer unit owns a spare electron. These spares are called pye electrons. They can hang out anywhere in their own orbital home. These electrons get fired with energy and this energy cascades away and is used in work. This is called electron resonance and is the basis of all biological clocks of the circadian rhythm. Like the sea, the sine wave of light traveling with a full load of energy has different impacts on molecular pathways, especially on those polymers that resonate.

Trees have perfected the photoreception of sunlight. This is what makes them the most special species on this planet. Trees receive quanta of sunlight and transform this energy into thermodynamic reactions. All of science cannot replicate the astounding feat of ingenuity found in a single leaf. Trees have a unique holistic chemistry. Their entire skeletal structure is composed of aromatic hydrocarbons linked in polymeric chains of extraordinary length. This makes the entire tree a photoreceptor. The aromatic structure of the polymers absorbs the energy of the electrons. The incoming electrons, as photons, join with other more normal electrons and add energy. The tree bleeds this extra energy off into growth. In addition, polymers in trees also occur next to one another in a form of columniation, long vertical arrangements of molecules. Sunlight traveling through such arrangements is capable of jumping quantum states, much as ordinary light can be energized into a laser form. The natural system of electron capture that the tree uses to grow may well have extraordinary ramifications for the human family that is running out of energy. A recent development in physics called the Bose-Einstein equation has demonstrated the plasticity of light energy that can move into different states and produce energy as various laser lights. Such a system might well be copied from the pattern of chemistry that exists in trees. This blending of the cutting edge of physics and polymers might well power the planet.

Despite the fact that sunlight is simple, it is, like water, not entirely understood. And recently a brand-new hybrid harmony has been demonstrated to exist. One photon of sunlight can react with one atom of matter, any matter. The photon and the atom unite to form a hybrid artificial atom. This hybrid is able to receive another photon, but this time it is a high-energy microwave. This unit cascades energy in the form of extra-energized photons. These can be captured. Trees may have been doing this for a long time.

The trees of the global forest capture the simplicity of the sun. They make food in a thermodynamic process capturing the primary resource of carbon. This process we do not understand. And yet, like fools, we cut them down. . . .

Sacred Trees Are the Riches of a Solvent Universe

SACRED TREES

In the global garden some trees are treated as being sacred. A deep reverence has emblazoned their image on the landscape. These holy events have taken place since the beginning of human time.

Sacred trees hold a message in silence. It is a stream of consciousness that can be tapped. It is not unlike the quietness of the painter and the stillness inside the notes of the composer. It is a sympathy with something grand outside of the human fold, a voice that transcends time and is heard down into the marrow of the bones.

Sacred trees are teachers. They enter and become the melody of the mind in passages of remembrance. They are a landscape in a hidden world. The world which will receive no physical human entry, only that which is invisible. The mind and its fruits of meditation flow into and out of these trees. That is why they were sacred and remain so.

The oak, *an Dair*, was sacred to the Druids. The oak held the legend of speech and could communicate with the heavens through lightning. Through their *Dair* the Druids believed that they could control the skies. They gave striking presentations from sacred wooded sites like Kildare in Ireland. Such places of ancient ceremony still hold on to the name rich in meaning. The name itself comes through the mists of time, for Kildare in very old Irish is two words: *Kil*, derived from *cill* which means church or consecrated ground, and dare, *dair*, which means oak, and is also the second letter of the ogham alphabet. The ogham script was a fifth-century Irish alphabet of twenty letters represented by a notch system for vowels and lines for consonants. This daily telegraph of soft-spoken words was chiseled into stonework all over ancient Ireland as tidings of goodwill and news of the past. This was how the names of these sacred places endured.

It was under a particularly large fig tree called a *pippala* or bo-tree, *Ficus religiosa*, that the Buddha himself received his first divine thoughts of revelation. And so from this tree with its shiny deep green leaves and soft malleable trunk with its yearly crop of *pippala*, delicious green fresh figs, a strikingly gentle religion was born. The tender young leaves, *pallava*, are used as a pot herb. The *Ficus religiosa* is the oldest sacred plant in India. To many in that country, it is the king of plants.

In California an eminent Cherokee chieftain and scholar, Sequoya, is remembered through his namesakes, the redwood, *Sequoia sempervirens*, and also the giant sequoia, *Sequoiadendron giganteum*. John Muir called the giant sequoia the "King of all the conifers of the world." These trees are the remnants of an ancient race of trees, which flourished as far north as the Arctic. The sequoias are the last green thread of history to a living Christ.

They seem to hold that story tall as they flood their green canopy into the azure skies. These trees offer a sacred bond to all who bear witness to their sentient personalities. The sequoias are sacred trees even now, to much of the modern world.

Another relative of the sequoia made itself known. It rose recently from the dead about fifty years ago. This tree was a water fir that grew in the garden of a village temple in eastern Sichuan, China. This lost relative turned out to be a sacred living fossil that had previously grown in luxury in North America and Europe three million years ago. This sacred tree was found to be a dawn redwood, *Metasequoia glyptostroboides*. Its sacred character to the simple villagers who revered this tree over the centuries saved it from extinction.

There are trees of sacred legend in North America. These trees arose naturally over time and produced such remarkable characteristics that they became sacred. Then a necklace of prophecy was placed upon them from the dreams of the aboriginal peoples. These were such potent dreams that they also helped forge a culture of respect for nature. Many of these trees were golden colored and botanically named *aurea*. They were a form of albino tree that defied the normal rules and regulations of the green world. Such trees arise as a genetic sport. They are unusual and rare. To the aboriginal peoples such trees are sacred. They carry signs and portents of life to come.

Another sacred tree is the sacred tree of the First Nations. It is a wafer ash, *Ptelea trifoliata*. This remarkable tree is a member of the citrus family, *Rutaceae*. It was spared for some reason by the last ice age in North America. By the law of averages, this tree should have been extirpated by the ice and cold. But somehow it survived. It also managed to get through the felling of the forests arising from the British Land Grant Acts. It is now endangered because of gross ignorance and cutting.

The wafer ash is a sacred tree, partially because it is a remarkable medicinal tree. It is a tree that holds a physic. Such trees have a war chest of chemicals at their disposal. These chemicals, because they were born out of the margins of life, do remarkable things. They can re-regulate the basic metabolic rate of the major organs and recondition them back to a normal, healthy state. Wafer ash has one biochemical, marmesin, that acts as a piggyback compound. It acts synergistically with other medicines, from aspirins to chemotherapeutic agents, making them more active and reactive. This was just one of the sacred aspects of the wafer ash used by aboriginal cultures over the millennia.

Sacred trees and sacred things do not have a place in a consumer society. They lie outside of the pale because there is no financial value in them. They are of the soul, transparent, to be forever treasured.

The Forest Food Line Is Being Ignored

FOREST FOOD

Back up to that sign. It says, "no farmers, no food, no future." It is a simple wooden sign, it should be seen on every roadside of the planet. The farmers are suffering. The people who are getting their hands dirty are not seeing the money for their produce. Farming has changed from a vocation into an industry. We are eating industrial food. And the seasonal delight has gone out of our food into the claptrap of convenience, convenience foods.

Planet earth is a sustainable unit. It is an orbital bell jar chemically contained by its own atmosphere. Photons from the sun pierce that atmosphere. This is the basis of all life. Life as we know it is interconnected in a vast pattern of living creatures, whose death breathes a further life into the sustainable diaphragm of that simple living. "We take and we give back" was the ancient credo of the wise.

The global garden provides our food. The food itself changes from habitat to habitat all over the world. In some places it is the grass family *Gramineae* that provides cereal crops with their enlarged endosperms that have been carefully cultivated with time. Elsewhere a tuber or rhizome is found in the wild, filled with complex starches and sugars. These tuberous roots produce potatoes and yams. From the sweet potato tree, *Manihot esculanta*, cassava or tapioca starch is obtained. Then there is the multitude of food from trees themselves, apples, oranges, pears, and nuts. All in all, the global garden offers a cornucopia of 80,000 species of plants as food.

But the eyes of the industrial food magnates only focus on 20 species of plants. These have become our global food source in recent times. Convenience has scaled these down even further today. Eight plants remain out of the 80,000 possibilities. The shoulders of the world rest on only eight food plants as a source of sustenance. The oral knowledge of the other 79,992 is rapidly being lost for future generations.

The famous eight are wheat, rice, corn, potato, barley, cassava, sweet potato, and soy beans. These are pressed and expressed. They are dressed in packaging and gilded for sale. They are colored and styled. They are beautified for the claws of the industrial market. They are emptied of all real nutrition and force fed to people who lust for their added sugar, salt, and fat. The modified foods are so easily digested that they are toxic to the working body.

The food from the global forests has been forgotten in the race to urbanization. The trees that were once called the antifamine feeders of the globe are being ignored or, worse still, forgotten. Many are cut down and some are slipping to extinction. Their fruits and nuts hold first-class protein filled with the essential amino acids for body building for true health. They have the three essential fats necessary for nerve functioning, brain building, and maturation. Both nuts and fruits hold sugars in a complex form that is sometimes polymerized for a tighter packaging within the forest food.

All of the natural foods of the global forest are complex molecules, bonded and bound together. They are slowly plied apart in the process of digestion. Their sugars are gentle on the pancreas. They enable the hormone insulin to work more slowly and carefully on its target sites in the body. These forest foods shield the body from diabetes. They have a cardiotonic function and they revitalize the brain.

The global forest has furnished these foods in the past. It is capable of doing it again. For the most part these food-producing trees such as the walnut, oak, pine, pea, ginkgo, mulberry, rose, beech, and custard apple families, among others, have not been genetically modified and hopefully they never will be. However, they have not been systematically searched and selected for natural hybridization to induce hybrid vigor for larger crops of bigger and better nuts and fruits. In the past, genetic sprouts that occasionally occurred with new traits were noted and some were even preserved. But for the most part these trees are being ignored and, even worse, lost by ignorance of them. There are no full collections of these trees in botanical gardens or in arboreta across the world. Neither the finances nor the will are in place for their absolute protection.

The farmers' sign has a bearing for the future of the famous eight. With the exception of cassava, the other seven have an Achilles' heel that worked in the agricultural past but is now under threat. It is their open-handed process of fertilization itself. They do not fare well with the increase in temperature and UV radiation which comes hand in hand with climate change. The sun's strength that fires growth will damage their exposed male reproductive structures. The seven species depend on a successful annual pollination. In all cases the male genetic parcel is dependent upon a generative cell. This cell produces a tube down which the male germ DNA travels. The tube stretches to window thickness through which the sun shines. In this exposed position the DNA is damaged. A cripple moves toward its future spouse. The saga of the crop holds its own ending.

And indeed the farmers' sign could be shortened to "no food, no future." End of story.

A Hedgerow Is a Living Chain of Forest

HEDGEROW HEAVEN

 The hedgerow is the living continuation of the forest. In some instances it is all that is left of the forest, a ghostly trail of tree stumps that invite native invasions from surrounding plants. In other cases it is the boulders and stones collected from the last glacial visitation, weathering slowly in the collection of trees and shrubs that have returned. In North America these places of collection are known as fencerows. There are more fencerows in America than in the island of England, which spawned the word and its use as an art form with living stakes of cut hawthorn called trouse, woven with long, flexible rods called ethers. This process was and is supervised by an official called a hayward.

The hedgerow is a corridor of life that expands into and acts as a boundary for farmers' fields. The hedgerow is seen wherever there is agriculture. The fields may be small and compact, or they may encompass large tracts of land as seems to be the modern propensity. Hedgerows have been reservoirs of great beauty and learning for little children whose first flowers were picked from these insular places. They rate with wetlands, open pasture, and connecting forests in producing food diversity.

On the faces of continental landmasses, hedgerows have great importance. Migrating songbirds, butterflies, and larger-winged creatures use them for resting, perching, preening, and as feeding stations in both the spring and again in the fall. They provide height for field predation and safety for songbird nesting.

Hedgerows have general plant populations that can be identified by related species from continent to continent. These places abound in grasses, sedges, and climbing species. There are mosses, liverworts, and fungi which intermingle with a variety of local ferns. There are shade- and sun-loving annuals. Biennials can also be found. There is a generous run of rugged all-purpose perennials. Both evergreen and deciduous shrubs and trees of every description may be found in the hedgerow.

Each plant species of the hedgerow has about forty species of insect in attendance on it. This is the biodiversity of the hedgerow and the insect species are predominantly beneficial. They in turn amplify other native life up to the songbirds and butterflies. Many small and larger mammals move into this area of plenty and the raptors float by too. This symphony of biodiversity also takes the annual monocrop of the farmer's field into its balance of predation and prey.

In the past handful of decades something else has been happening in the farmer's life. He has been pushed to use every square inch of field to produce more crops. The low price of food required the farmer to do more with less. This demanding situation has forced farms to remove their hedgerows. They are coming down all over the North American continent, in Europe and also in Asia.

The fields now have no boundaries. They are vast tracts of land like the ocean. They have no end. The empty field's soil surfaces move in shoals of erosion in the spring rains and form a pattern of eternity with the summer crops. There are no places left for cover and for perching for songbirds, for butterflies, for mammals, for any of the great arena of biodiverse insect life to call home.

Removal of hedgerows has given greater efficiency to the use of pesticides on crops. All agricultural pesticides are killing compounds designed for use on the food crop. They may also have the ability to move internally inside the food and become systemic poisons opening the way for pathogenic bacteria to live on the crop.

They may be genetically modified in situ either in the seed or in the plant as it grows. Pesticides are toxic killing compounds that spawn daughter killers and act synergistically to turn others into even more toxic aerosols. These pesticide aerosols travel great distances and concentrate at the poles. Thus it is so, that from these vast tracts of famine comes the sorrow of our food.

The removal of hedgerows, the opening of fields, and also the move away from sustainable farming practices is doing something unprecedented to nature. The soil is being fertilized with artificial fertilizers that add nitrogen in the form of water-soluble nitrates and nitrites. The excess of these salts dissolves in the water columns of the soil with the spring applications of fertilizer over gigantic fields that leach directly into streams and then into local watersheds which drain into rivers that reach for the sea. Finally they spill into the bays and gulfs of the great oceans.

Hedgerows are part of a pattern of sustainable thinking. They were the part of the visible vision of the countryside that became the landscape of life. Their existence measured the success of a farm and nobody knew it. Their presence involved the biodiversity of life whose richness gave a pleasure to daily living. But farmers have to eat too. They must be paid a fair price for the food that they produce. They cannot and should not feel that the last inch of a field should go into production to benefit a middleman. That last inch affects the oceans and the fish that depend on oxygenated water for their lives.

A hedgerow is the living continuation of the forest and yes, it is a symbol of life, all life, it would seem.

A Warning from the Forest:
"Bacteriophages and Blue-green Algae
at Work," Do Not Disturb

WARNING FROM THE FOREST

🌿 Once upon a time there was an island called Guam. This little island was surrounded by the sky-blue waters of the Pacific Ocean. On the island there lived a race of aboriginal people. They were called the Chamorro. These peoples hunted and fished the great waters. They celebrated life with their festivals from year to year. They were a healthy people.

The island of Guam is forested. The trees are those common to the old world and are part of an evolutionary partnership that gave rise to more advanced trees like the oaks and hickories of the new world. These trees are members of the *Cycadaceae* family. They stand about twenty feet (6.5 meters) tall and have a tight topknot of palm-shaped leaves that are deep green and hide a cache of dark brown shiny nuts that are sometimes the size of a walnut. These mature nuts are much prized by the creatures of the island of Guam. These trees, too, are used as food and carry common names of bread palm and funeral palm. They also produce an edible, somewhat starchy and sticky flour called sago. The nut flour is the basis of the flat bread called tortillas that accompany most meals.

Once a year the Chamorro skipped the tortillas, substituting meat in honor of one of their festivals. The meat they put on their plates was somewhat unusual—the meat of a bat. It was a particularly large juicy bat that went by the name of *Pteropus mariannus*. Because of the cute profile of this bat, it had a common name of the flying fox. These bats were cooked and enjoyed as the number one gourmet item on the Chamorro menu.

In the normal run of a Chamorro's life these bats were difficult to catch. They were elusive creatures that darted in and out of the cycads in the dead of night. They were difficult to see and even worse, more difficult to snag, and when a carcass was brought home they were doubly enjoyed as a once-in-a-year luxury. Each morsel was chewed with sheer delight down to the brittle bones.

In due course, the peace of this pacific paradise was broken. War broke out. It was the elastic war of World War II that clung like a limpet to all that was beautiful. The Chamorro people looked with envy to their new role models. Slowly they acquired the guns they admired. And soon the bats were daily fare. Until they were extirpated, of course.

In a fairly short time after the windfall of guns, sickness arrived. Almost to the last man the Chamorro began to show startling changes. These were of a neurodegenerative disorder. This disease had not been seen before. Traits of Parkinson's disease, Alzheimer's, and the dreaded Lou Gehrig's began to show their deadly trail. People began to die. Not a single one of the deaths was of the customary peaceful passing away representative of this particular paradise.

It would appear from the detective work of scientific and medical investigation that a trail of evidence led to the poor innocent bat called the flying fox. The bat ate the fruit of the cycad. The cycad in turn protected the fruit with a good dose of poison. This poison happened to be a neurotoxin. It was a simple readjustment of a nonessential amino acid called alanine into a raging neurotoxin called beta-methyl-amino-l-alanine. This neurotoxin accumulates in the body. And it comes from a very common source.

The cycads, like many vascular plants, live in a green companionship with all kinds of other plants. This quiet mutually beneficial behavior is more common around the food market of the tree's roots. In this fertile region a large amount of trading takes place. In the case of the cycad, the trading is with a group of soil-borne blue-green algae that offer nitrogen to the tree. These ground huggers fix nitrogen out of the air with their very able nitrogenous enzyme systems. The barter is then offered to the tree. In exchange for nitrogen which has been fitted into the converted amino acid alanine, the blue-green algae then gets its fix of sugars. The alanine is water soluble. It is quickly transported up the phloem system of the tree into the nuts for their greater good.

The cycad nuts that had been ground for flour to make the tortillas had been previously washed in many changes of water prior to grinding into flour. These washings are common in food preparation of many indigenous comestibles of aboriginal peoples worldwide. This worked for the cycads, but the bats boiled in coconut milk are something else. They carried a full load of neurotoxin. When these were eaten on a daily basis the toxin began to accumulate in the eaters' brains. And . . . the inevitable funerals took place.

But there is a postscript to this story. Blue-green algae are common appendages to vascular plants worldwide. These algae live in the soil, all soils of the global garden. These algae are in turn regulated by bacteriophages, the most numerous species on earth. These are also the most genetically active. Vascular plants need nitrogen to grow and these algae supply this essential commodity. Artificial genetic modification of vascular plants like the crop plants of our food might alter this exchange of nitrogen and a benign chemical like an amino acid might be turned into a raging killer. And then we would all be in the soup. Coconut milk, of course. This is more than just a passing thought!

Trees Communicate by Infrasound

SILENT SOUND

There is a symphony of sound going on around us every day. It doesn't matter where we live or who we are, the sounds are there. We hear them and maybe even recognize the song, the grunt, or the small groan. They are part of our lives, part of the symphony of sounds that bathe the moments of our daily living. These are the sounds that we both hear and register.

There are other sounds out there, too. These are the portions of the sound spectrum that are inaudible to us but can be heard and registered by many other creatures on this planet. One of these portions are all the sounds of twenty hertz and below. These sounds produce long, stretched-out sound waves that travel and can be sensed across immense distances.

Even planet earth rings a bell of life as the planet moves ahead in its orbital path. The matter of space is forced into a giant wave pattern that travels outward into the galaxy in a registration of silent sound. Earth rings our bell of life for infinity to behold. The extreme sound, that silent sound that is produced, is known as infrasound.

This silent sound called infrasound is produced by a great many things, some of them known and identified, others remaining elusive so far. The sound waves of infrasound are like ordinary audible sounds, but as they travel through air, the lower their frequency, the greater the distance these sound waves will travel. Volcanoes and tectonic plate movements produce silent sound waves. Events like hurricanes and thunderstorms generate their own infrasound symphony and so do many of earth's larger creatures like rhinoceroses, elephants, and whales. So do trees and forests of the plant kingdom because of their great size and tensile movement in air. Within the forests some birds can hear silent sounds, too.

Sounds, all audible sounds, are a means of communication. The communicator sings a song and the pattern language of the brain fires out the messages of its meaning. The listener listens and understands in the common ground shared between the species' brains. Elephants, for example, generate silent sound speech of fourteen hertz. This frequency is low and can travel a long way in the ground, even through forests and grasslands, to another listening elephant that hears the waves as resonance, leaning the head and trunk forward into the sound for its registration.

Honeybees, too, can communicate with infrasound. The hive, when it is at home and is almost complete in numbers, acts like a warm-bodied animal. The bees of the hive hear the silent sound generated by hurricanes, twisters, thunderstorms, or tornados. They move into a foaming behavior somewhat like swarming. They change the frequency of their own sounds to a higher pitch. This alarm signal in turn generates the increased production of venom for hive protection. All bees become killer bees under these circumstances for protection of the hive. The young embryonic brood it contains is paramount.

Infrasound is also heard by some people, and it is registered as an intense emotional experience. This experience is quite often felt in the cathedral of the forest where the low-frequency sound waves are amplified by the musical instrument of the human body, which is built like a large cello. The ribs and their intercostal movement in breathing are confined by the neck and again below this musical case by the soundings of the dividing diaphragm. The low-frequency amplification affects the sympathetic nervous system, where it has a stabilizing effect, and it also appears to affect the emotional centers of the brain. The acoustical experience of people who can register silent sound as a form of cross-species communication testifies to its being felt primarily in the chest as a reaction of immense intensity together with a tightness that can bring one close to tears. These kinds of feelings are commonly felt listening to music all over the world.

All trees of the global forest produce a fingerprint of sound. This sound is as individual as the iris is to the eye or the thumbprint to the hand. The morphology of each tree with its trunk or trunks and the pattern of its canopy composed of leaves with their unique shape make for a sound that is unique to its generator. The sounds—both audible and inaudible—of the pine are sharper when compared to the rounder sounds of a red oak. The movement of air as it travels through and with the pine is more finely dissected by pine needles to make the sharp whine, while the red oak leaves are more like the flapping of sails on a yacht.

Infrasound might well be the way by which the trees of a forest communicate with one another, like a colony of bees. Warning calls of infrasound might connect one tree to another in a dull roar of low-frequency sound, which is heard by children and many people and is registered by a deep choking if not a strangling emotion on the cutting down of a particularly large tree or an expanse of forest.

The elders of the aboriginal peoples of the Canadian rain forest of British Columbia describe this sound as "the weeping of the trees. . . ."

Medicine Is Found in the Wood of Many Trees

MEDICINAL WOOD

The inhabitants of the ancient world had a knowledge of trees, timbers, and wood. This came in part from a natural curiosity about the immense biodiversity around them. It came, too, from ritual dreams. This knowledge was collected and shared as part and parcel of the collected works of their oral traditions. From this foundation of learning a base of working knowledge was laid. This knowledge gave them the power to survive as a people.

Each species of tree produces its own identifiable wood and timber. And within the species itself there is variation depending on the tree's habitat. Variation is also caused by internal endophytic fungi which live within a tree's anatomy. And even more and stranger changes are caused by hormonal flux within the tree that may not be altogether normal in character, but is reflected in the tree's wood and timbers.

The wood of a tree is a mixture of natural polymers, one of which is lignin. Lignin is laid down within each cell of the tree, in the cell wall, in a matrix, to give the tree strength and the ability to stand tall. Decisions are made by the tree to tighten this lignin matrix by cross-bonding for additional strength in tissues called collenchyma and sclerenchyma. These bonelike sections of the tree's anatomy provide for branching and canopy strength and give rise to the tree's unique physiology or growing form.

Outside of the skeleton itself, a tree manufactures a large range of polymers. These long chains of chemistry are ideal for a growing tree and can be added to when needed. The chains can be tannins, complex sugars, mucilage, and the sound attenuator suberin, or cork. There are many more important polymers like resins and waxes. In addition, there are a host of aromatic biochemicals that move in and about the traffic of the cambial skin of the tree, some of which are known and many yet to be discovered. There are the aerosols, the elusive chemical messengers, within and released into the atmosphere by trees and forests. These fragile molecules have short half-lives and change as they move and are released into air, water, or soil. In addition to aerosols, there are the chemical children of these parent compounds, waiting to be switched into some other quantum state.

The magic of medicine is found in this mixed chemistry of wood. Every tree in every forest of the global garden is different. The medicinal chemistry changes from one year to another and within the seasons, too. Sometimes there is an induction of these chemicals in one tree brought on by the stress of disease or weather. Within a short period of time, the nearby trees of the same species are found to increase these same chemicals, as if there were a telepathy between all the species.

There is wood that is distinctly medicinal and other wood that is poisonous. All members of the *Juglandaceae* or walnut family in the global garden produce wood that is medicinal. And most members of the *Leguminosae* or pea family, as in the *Gleditsia* species, produce wood that is poisonous to termites. There are families that produce fungicidal aerosols such as the *Cupressaceae* or cypress family. These were used for sweat lodge cleansing and on the Indian subcontinent for funeral pyres as after-the-fact air fresheners from death. There are woods that are antiviral in the *Betulaceae* or birch family. Aerosols from these trees would benefit prostate cancer, kidney dialysis, and organ transplant patients. There are woods of the same family that are infected with a fungus that persuades the *Corylus* species to produce taxanes, an anticancer drug of immense importance in breast cancer and other reproductive cancer treatment.

All medicinal wood of the trees of the global garden release medically active aerosols in minute amounts. Once these chemicals are airborne they can travel great distances. They can hitch a ride on a pollen grain for short distances or they can travel on micron-sized particles. The aerosols have the ability to disperse in airborne molecular traffic. Here they may release themselves as potent antibiotic, fungicidal, or anticarcinogenic compounds. They can act as an antiviral, antibacterial, or antifungal shield in the air from the tropics to the poles.

Medicinal wood functions in the home in a similar manner. Furniture, fixtures, or even the house itself may be built from medicinal woods. These gas off minute amounts of protective chemicals that travel in the airways of the house. The vortex of them would be around the washing machine or dishwasher, which mixes humid air into the closed environment and solubilizes the aerosols for further movement. These biochemicals are beneficial for household health.

Indeed, the ancient world had a vast knowledge of trees, timber, and wood. The aboriginal peoples of that world even had a medicine for bereavement, a concept foreign today. They burned branches of *Pinus strobus*, white pine, in a house of terminal sickness or death. The fungicidal white smoke was for the living to help them repair their broken lives. It was thought that this particular smoke had a healing quality for the spirit, too. Surely it has.

Trees Define the
New Sexual Revolution

THE SEXUAL REVOLUTION

Trees copulate in copious amounts. Plants have had the good fortune of being outside of the rigors of religion, so they do as they please, when they please, and much more important, how they please. They are plants after all and everything goes in the plant kingdom.

The life of a plant or, indeed, the life of an animal is bordered by birth and death. These two events are constants in the currents of life that flow through the channels of procreation. All of life from the plant to the pigeon carries the burden of replication down the stream toward death. Some species do it better than others. Many produce flowers, mushrooms, calyptras, or conceptacles; others, gels, perfumes, petticoats, or even pipes. All replicate to give birth in the dance of life, for replication is of itself the rather measly definition of life, as we all know it.

For a plant such as a tree, sexual parameters are paramount to ensure a continuation of life. And there is diversity within these parameters. This is true biodiversity, an expression of genetic flexibility, which gives all creatures, all plants and animals, the edge. Without this cutting edge of change, life would stagnate. A redundancy would set in and the identity of the individual would be lost within each species.

Heterosexuality is the more normal process for a tree—the mobile male impregnates the immobile female ovum. The two new gene sequences line up for a happy marriage and aim to live forever in that state. Trees and forests appreciate heterosexuality because this keeps the nuts, the seeds, and the samaras on track for family life.

However, homosexuality also exists in the forest. Sometimes it is part of the normal family fare and other times it is expressed as a stress factor when nothing else works. But nonetheless, homosexuality for a tree is a valid and natural means of sexual procreation. Certain species of trees can jump the normal barriers of reproduction and can produce an embryo from a sexual act of itself, be it male or female. The female process is called parthenogenesis. The male process is unnamed but exists in the denuded areas of North Africa and possibly elsewhere.

Homosexuality is an expression of biodiversity within sexuality. This expression is not uncommon in nature. It is a means of giving the genetic code the ability to produce sports. A sport mutant has a genome that can produce enzymes for change such as greater musical skills seen in the arena of the composers or a golden or albino *forma* in the plant kingdom. Occasionally such a change can produce a genius. A genius almost always reroutes cultural philosophy.

In addition to heterosexual and homosexual reproduction, trees can reproduce by vegetative means much like the newly discovered stem cell. Vegetative reproduction is not a simple molecular process. It is built into the cell. In the typical case a nonreproductive cell, an ordinary somatic cell that carries a run-of-the-mill diploid chromosome count, can explode into a species that is a molecular mirror of itself. As in cloning, nothing new is learned of sexual biodiversity except more of the same in an identical pattern of life.

Trees can divide up the family household into the sexes, where the female followers are designated a tree of their own. These are female trees. Males are moved out into separate male or pollen-bearing trees. In this dioecious style of life, polygamy is seen in more numerous male trees. The female species is more elite and seen less often in the forest.

In other situations the male might be out on one limb and the female on another. The tree is quite often in a windy situation and the pollen is brought to the female on airstreams or by flying insects or birds.

There is also a tighter family unit where the male and female live in close conjugal comfort in the same flower. The couch potato pollen tends to be a little heavier than his windblown cousins while he sits and waits for his close sexual call. This seduction works for many common tree species too and is called monoecious behavior.

In the sexual life of a tree there is the more delicate question of timing. Sometimes the pollen is ready too early, other times the beleaguered pollen is too late. Occasionally the boy next door pops in while the pollen sulks. And again the ovum may pout until something better comes along on the wind of change to fertilize the egg.

In addition, there is the possibility of another form of sexual behavior in the forest. The whole tree itself may represent one partner in a sexual relationship with a fungus. The lifeline view of this pairing is a long one and ends only in the death of the tree itself, which in turn initiates sexuality in the fungus partner who begins another tryst as a basidiospore or an ascospore.

Only sex spells true biodiversity. And every sexual revolution carries some cunning for another sexual leap of any species on the road to perfection.

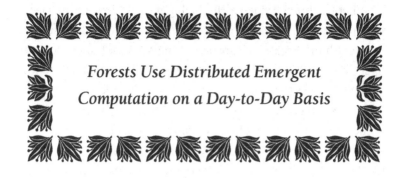

*Forests Use Distributed Emergent
Computation on a Day-to-Day Basis*

COMPUTER COUNTRY

Pouting painted lips are part of the pop art image of the human race. A mouth with lipstick kisses the pages of women's magazines and talks its way into newspapers. Every face has one mouth and that mouth has a function. It opens the passageway to the two pink sacs called lungs and keeps them filled with air; hot air, that delicate specialty of the human face.

Trees are different. They do not have one big mouth. A deciduous tree has billions of them. A leaf surface is covered with millions of mouths. These are tiny cavities that open and close just like the human airway. The green lips of the mouth of a tree are called a stoma and are made up of two long cells. The mechanism for opening and closing is performed by another two smaller cells on either side. These are called guard cells. When the tree wants to breathe, the stomata are open. When the tree wishes to stop breathing the stomata are closed. The guard cells go flaccid to close and turgid to open, again.

119

What the human does with one big mouth, the tree manages with millions of multimicron-sized mouths. Small is beautiful for the mouth of a tree. The function is the same. A human mouth opens up and draws in oxygen. This gas fires the burning of food, which produces the usual toxic gas, carbon dioxide, of which the body must rid itself. This exhaust gas of combustion is pushed out of the lungs in the form of an exhalation of breath.

On the other hand, the tree needs the exhaust gas, carbon dioxide, for body building. So the tree opens its stomatal mouths and breathes carbon dioxide into the stomatal space. The tree reworks the carbon dioxide into carbon-bearing sugars and flushes out a nontoxic gas called oxygen. This gas blends immediately with the surrounding air, enriching it.

So man and trees are caught in an interconnected cycle. The man produces the carbon dioxide gas that the tree desires and the tree produces the oxygen gas that man desires. They are both part of this cycle of need. On closer examination both are a bit hairy, too. A human mouth is sometimes surrounded by a great deal of hair. A tree's can be, too. In both tree and man the mustache protects the mouth from the elements. But the tree has taken this one stage further. The tree uses a million hairy mouths for reducing the need for water in arid conditions. The moisture of breath condenses on the hairs as water vapor and reduces the total water loss by the tree.

In the deep centers, or limbic regions of the human brain, the process of breathing is regulated. This automation is controlled by the computerlike system of the brain, a gene-derived binomial function of command and obey. The trees do this, too, in their own fashion. They have a large leaf surface sitting in the tree's canopy. Not all of this area is needed in synchronous breathing. In some areas of canopy where the sun is a bit too strong, the stomatal mouths shut down, and to get their breath other, more shaded regions take over. This is regulated by some deep center also.

It is achieved by distributed emergent computation, a process which is also common in other complex systems such as the immune and nervous systems. This kind of household management is also found in large ant colonies, in congregations of yeasts, and in slime molds where decisions based on food supply of gases such as carbon dioxide, oxygen, or nutrients rule the roost.

It is not exactly known how the trees and all vascular plants regulate their breathing or respiration processes by distributed computation. The control centers, if they exist at all in the trees, are difficult to locate. They might be companion cells or hormonally triggered.

The transpiration process in trees, that is the way trees breathe out oxygen, water vapor, and other aerosol chemicals into the surrounding air mass, is not understood either. It seems to be part of the arena of this fluctuating form of computation.

At the moment the best guess, and it is only a guess, is that the underlying principle is a mathematical one called cellular automata. It is thought that cellular automata may be the underpinning for all phenomena in life including the description of elementary particles and even intelligence itself. But science is a long way from understanding the simplicity of such complex forms, let alone the mathematics of them. Cellular automation is seen in slime molds, myxomycetes, yeasts, and fungi which exist separately as individual cells. Then, as a unit, they decide to get together. The yeast will form a torula or resting stage where the cells unify and can exist indefinitely. This cellular communication is not understood and has not been studied.

In the meantime, trees with their so-called lack of intelligence continue to sequester carbon dioxide gas for human life. They continue to filter water in the phloem and sieve tubes of their plumbing. The water coming out is certainly cleaner than the water going into the tree. Groundwater that is moved into and evaporated from the millions of stomata gives a new meaning to water cleanliness. And the gas called oxygen, rolled out of every stomatal forest pore, is recycled into our communal air bag called the atmosphere. That should give a new meaning to gratitude.

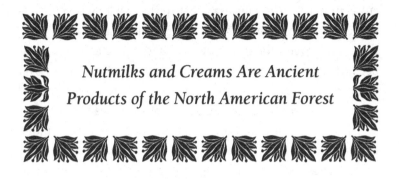

*Nutmilks and Creams Are Ancient
Products of the North American Forest*

NUTMILKS AND CREAMS

🌿 In the ancient world of North America there was a daily special. It was on offer during the winter months. The special was a milk. It was on the menu with wild duck, deer, and hare. It was added as a gravy and thickened with the mucilage of corn silks. These golden threads from mature corn husks were stored and saved for the sweet flavor they imparted into a meal. The milk was no ordinary milk, it was a nutmilk.

Across the world in Asia another milk product was discovered. This milk was extracted from beans. So it, too, was of plant origin. It was soy milk. But the Asian milk was taken one step further and made into a soft vegetable cheese. The milk proteins were precipitated out of a solution using either dilute acids such as rice wine vinegar or magnesium chloride. The proteins were strained, dried, and caked. The cheese called tofu was made.

The North American nutmilk has never had a modern name, much less an identity as a food. It was an important native food of the aboriginal peoples. Like maple syrup, a custom of collection arose around it. The milk was kept fresh for a long time. It was stored in birchbark *makuks*. These were folded, not unlike briefcases. They were stored in movable buildings. The nutmilk was used during the wintertime in a slow meltdown of flavors over an open fire and many a testimonial was declared by astonished pioneers, afterward, to the fine quality of such meals.

The North American nutmilk was made from various indigenous nutmeats. These were primarily nuts of the hickory or *Carya* genus. Large quantities of nuts were gathered from huge trees. These were *Carya ovata*, shagbarks, and *C. laciniosa*, kingnuts. The nuts were collected in the fall. They were dehusked by hand. The unshelled nuts were dehydrated in open ventilation. This process dried down the internal intima covering and the testa meat inside. With the shrinkage that came with this, extraction of the meat inside was easier. Some of these nuts were used for finger eating. Others were broken open in such a way that the meats inside were not fragmented.

Then the hickory nuts were put through a hot water extraction process. A birchbark kettle of water was taken to a rolling boil. The smashed nuts were thrown into the water. In minutes the protein and fats were extracted. The fats floated to the top of the kettle as an oily layer. This layer was skimmed off and strained through fine human hair strainers designed for this purpose. This became the nutmilk which was stored in *makuks*. Occasionally this milk was reheated and evaporated down further to a greater viscosity. A delicate nutcream was produced. The flavor became more concentrated with the heat treatment. The nutcreams, too, were stored in smaller *makuks* in conical larders for special use.

Another quite extraordinary product was made from the nutmilks and nutcreams. It was a wine of low alcoholic content. The creams and milks were exposed to wild yeasts, of which there are many, interesting and novel. One more northerly yeast is black, *Saccharomyces nigra*, and works in the cold temperatures in North America. The creams and milks had sufficient complex sugars to ensure that the fermentation process proceeded without delay. Again the strength of the wine depended on the power of the sun. In a good, dry, sun-filled summer the complex carbohydrates multiplied in numbers. This imparted a sweetness to the wine which, in turn, spelled out the wine's strength and keeping capacity. This creamy, delightful wine went by the aboriginal name of *pawcohiccora*. It was a choice beverage taken to while away the boredom of the long cold winter months in the company of family, friends, and, of course, stories.

Another species of hickory, *C. cordiformis*, known as the bitternut or pignut hickory, produced a fine grade of oil. This was extracted from a much smaller nut than in most other species of hickory, which in turn was produced in far greater numbers on these trees. The bitternut trees grew to immense sizes, sometimes surpassing 20 feet or 6.1 meters in circumference, with towering canopies. A large quantity of this hickory oil must have been necessary for daily life because it was a trading commodity with the incoming pioneers, who used it as a source of oil for lighting their lamps. Hickory oil was also applied as a hair pomade in certain rituals and was used in cooking. It had many medicinal properties as well. It is quite possible that the hickory oil was used in combination with the nutmilks and nutcreams in dishes that are now unknown to us. The weak link in any oral history is memory. And that has unfortunately been lost.

Many trees representing the hickory family still remain in North America. They are but a fragment of what was there in the past.

This family of trees died out with the last ice age in Europe. Only the pollen occasionally comes up in digs in Poland. The tree will not thrive in Europe to produce nut crops because the solar conditions have changed since the last ice age there.

All of the hickory family produces a particularly dense wood together with a colossal nut crop. This means that these species have a high demand for carbon in their growth. The hickory can sequester carbon out of the atmosphere like no other tree can. They have done this in the past in immense stretches of virgin forest and they can do it again.

For the farming community of eastern North America, all of the hickories including the more southern one, the pecan, *C. illinoensis*, and all of their hybrids, natural sports, cultivars, and crosses represent a new idea: nutmilk and nutcreams in food production. This idea is, by its very nature, twinned to the reduction of global warming by the carbon sequestration ability of the hickories. The two go hand-in-hand in logic to build a better and brighter future into this new millennium.

THE INVISIBLE FOREST

In the beginning was the world. This world was made of land and sea. It was good. In that sea there was something small, so very tiny that it would only be visible when massed together. Each had one green eye. This was a unicell.

In the seas of the world the unicell evolved into many plant life-forms. These became the landscapes of the oceans. And as the continents separated the sea plants moved up into tidal zones that were rich in nutritious foods. In time the unicell's facial features could be found in the brown, waving *Laminaria* forests, rich in their mucilaginous greatcoats of slime. They could be seen in the Sea of Sargasso, replete in later times with the life of an eel. The unicell turned face to become the deeper red-celled *Rhodophyta* plants of the ocean, clinking with silica armor plates. The unicells mingled with themselves too, and formed the ribbons of blue-green algae that could multiply by akinete and heterocystic sporulation.

These unicells and their companions in simplicity make up the phytoplankton forests of the ocean. To the human eye it is an invisible forest. They form a giant mass or swag of growth of biota that free-floats back and forth from continent to continent in the drag of the earth's orbital shift. They sometimes get shoved onto continental shelves by climatic turbulence. But mostly they follow the sun in slow motion, in its north-south trek and the warming trends that arise from these solar seasons.

The unicells can do something else. They can swim. They use flagella legs to kick themselves off in motion. Others use the protein slide of the snake. Either method of locomotion helps them to rise upward and downward in the columns of seawater which are their homes. The limit of their home life is the unicell's genetic size. The cells with the greatest gene number can navigate the greater depths, down to two hundred meters. The cells with the smallest genes stay at the top twenty meters or nearer the surface of the oceanic columns.

The life of the unicell is fraught with dangers. These tiny organisms can be taken over by monstrous gangs of bacteriophages. These viruses loot and pollute the unicell. They inject genes into the unicell genetic code and make the unicell into an absolute slave. Sometimes the unicell escapes using the salve of its own antibiotic. Other times the unicell loses all and goes back into the solution from which it has come.

This flood of protein and carbon is the real green grass of the ocean arena on which all creatures feed. It is thought that there is a 40 percent turnover of the unicell biota on a daily basis. This flush of nutrients activates cell division. Expansion is controlled by bacteriophages. This mass of life fuels the great carbon sequestration ability of the ocean.

The invisible forest of the unicell is almost equal to the land-locked forests of the global garden in its ability to flush oxygen into the atmosphere and to sequester carbon into a stable of organics from its gaseous form. The unicell is the most abundant photosynthetic system on earth. The unicell lives in the oceans, but it also lives and thrives on land.

The operation of a tiny unicell and a tree are similar. The unicell has a cache of chlorophyll and so does the tree. The housing of the chlorophyll in the unicell is a bit primitive when compared to that of the tree, but both do the same thing in the end. The chloroplast of the unicell sequesters carbon dioxide out of the atmosphere from its dissolved form as carbonate. The unicell chloroplast uses the energy of the sun to refit the carbon into carbon-rich sugars. The unicell does this in daylight. However, at night the unicell changes its tune. The cell fires up its nitrogenase enzyme to take up a little nitrogen for itself. It does this either from the water or from the air. And the unicell refits this nitrogen with a little carbon into protein. The unicell has a further flexibility, in that while the nitrogenase enzymes usually need iron as a catalyst, the unicell can bypass this in time of need and makes use of nickel or copper from the ocean's currents.

The unicells are also responsible for the algae flush or toxic red tides that are sometimes seen as killer tides around coastal shores. These occur in the heat of summer, either north or south depending on the season. They are triggered by a flush of excess nitrogen as nitrates or nitrites in river runoff from spring and early summer household and farming practices with artificial fertilizers. The nitrogen-rich waters take a little time to flush out into the ocean columns. The excess nitrogen activates the nightly nitrogenase activity of the unicell. The unicells grow and divide. Then they divide again. Soon there is a mass feast for bacteriophages. These bacteriophages bear down on the unicell flush but they need oxygen for digestion.

The water becomes oxygen deficient and hypoxic. The fish die. The ocean region dies too. Ocean repair is slow.

The invisible forests of the ocean produce about half of the world's atmospheric oxygen. The other half is produced by the great forests of the global garden. There is a synchronicity between the two forests of the planet, the tiny unicell and the massive tree. The font of this synchronicity is the interface of the ocean itself, where the water meets the air. At this point of mixing of solutions and gases something invisible occurs. The droughts incurred by cutting down the global forests make the surface of the ocean drier too. They make the sea solution saltier by evaporation. This changes the chemistry of the surface of the ocean and its ability to pick up carbon dioxide and convert it to carbonate. This starves the unicell and reduces atmospheric oxygen. And only One Being knew that the balance was so fine. "In the beginning was the world. . . ." So very, very fine.

Global Forests Control
Global Warming

GLOBAL WARMING

It is no wonder that tennis is popular. A singles game has been played since before Cambrian times. The sun and the earth lob balls of energy back and forth between one another through a net. This game has been going nonstop for a long, long time. There has been the odd hiccup with a stray meteorite or the occasional volcano doing its thing, but the game of tennis has never come to a grinding halt. Not until now. And future games don't look too bright either. Something has gone wrong.

The unknown quantity is the net itself. Like all other nets of various games, it sets the limits and lays down the boundaries of the game. Without the net the game itself cannot be played. It would have no meaning whatsoever. Therefore the net is important, very important. The net is earth's atmosphere.

The net has a story and it also has quite a unique history, too. Unlike her sister, Venus, and her brother, Mars, Mother Earth, our little gentle planet, can support life. The atmosphere can do it because it has a good supply of oxygen and an even better amount of nitrogen. These occur naturally on earth. Sometimes they are in a gaseous form; other times they are locked up in organic compounds. They, oxygen and nitrogen, are both set into a cycle of life that controls another atmospheric gas called carbon dioxide.

Carbon dioxide is a killer. It is known as a toxic gas to us and it is found on Venus and Mars in very high concentrations. The levels of carbon dioxide in these orbiting neighbors will not and probably have never supported life, at least as we understand it. And the very strange thing is, at one time, the earth was just like its neighbors with their toxic levels of carbon dioxide. The earth did not support life until something wonderful happened. The history of the atmosphere is tied into a Big Bang theory, any one of them. It is tied into evolution and a divine plan, any genesis.

The underlying weft in the loom of life is mathematics, elegant, simple mathematics, based on unity. This is what has touched planet earth and this is what has transformed it into the gem of the galaxy.

But the history of the atmosphere was not always so sweet. At one point in Precambrian times the atmosphere is believed to have contained 98 percent carbon dioxide. There was land and there was water. Into that water a kiss of life was offered. Whether it was methane or some other simple organic compound is a moot point. From that carbon kiss evolved a living form whose skeletal system was based on that carbon itself. From this kiss evolved bacteria, algae, lichens, fungi, jellyfish, and sponges. These creatures slogged along and eventually helped the great carbon period of earth to bloom. These times are known as the Carboniferous Period.

It was in the carbon-rich seas of the Carboniferous Period that life got seriously down to business. Biodiversity began to beget more biodiversity leading into the life-forms that are more common today. The atmospheric carbon dioxide gas was transformed into the carbon of carbonate shells and sea creatures in the warm oceans. They divided and set forth and multiplied in mathematical order with precision and grace. These creatures reduced the toxic carbon dioxide gas in the atmosphere and transformed it into the water-soluble carbonates of our ocean beds, our rocks, and our mountains. This carbonate enabled the atmosphere to increase in oxygen.

An atmospheric net placed itself between the sun and the earth. This net protected all life from ultraviolet radiation damage. It promoted the protection of the fragile life-forms of birds, mammals, and plants evolving on land. These, too, were based on carbon. As the giant ferns, the huge mosses and club mosses evolved along with trees like horsetails and even stranger cycads with their strap-leaved ponytails, the kingdom of the flower, of the tree, and of man was coming into the mathematical formula of unity. All of life was good; it hummed with an extraordinary biodiversity.

The game goes on, this tennis game between the sun and the earth. The sun lobs balls of heated high energy toward the earth. The planet still finds these lob balls too hot to touch so they are bounced to cool them down. The earth bounces them between the atmosphere and earth itself, and when they are cool enough much energy is used to fuel the planet while the rest escapes through the net and back out toward the sun.

This bouncing and energy reduction is called "the greenhouse effect." The short wavelength of the sun's radiation gets reduced down to a longer wavelength of less energy. This infrared bounce heats up the atmosphere and keeps it warm for life on land and sea. It is reflected back and forth, which amplifies its effect. This is good for everything. The game goes well.

In the last two hundred years, the carbon reserves have been and are being burned. This liberates carbon dioxide into the atmosphere. The great global forests that sequester or thirst for gaseous carbon dioxide have been cut down. Both they and the carboniferous creatures of the sea are the vital cogs in the carbon cycling wheel. The carbon dioxide is building up again in the atmosphere. It has doubled in these last two centuries. This new carbon dioxide in the atmospheric net bounces the infrared energy so hard that it increases the greenhouse effect. The atmosphere is heating up.

The game is coming to an end. There is still time to fix that net before the great umpire in the sky calls out for once and for all a cry of "Net ball. Game over! Please. Net ball. Anyone else out there for a game? There has got to be somebody . . . somewhere."

The Sun Produces Polarized Light and Nature Reads It

THE BIRDS AND THE BEES

Whether it be a scramjet at supersonic speeds or the lazy lope of a butterfly lolling in the winds, all flight needs lift. Birds are no exception to this rule and neither are bumblebees. Many plants need lift, too, for seed dispersal. Trees and forests supply lift and landing for their flying population. They supply airways and runways, too. Safe landings and takeoffs are as important to birds and insects, as is the vital matter of their food and water.

All trees throughout the global garden provide this lift and protection of airways for flight. Some trees advance a step further and grow thorns for additional comfort or cover. The airways that birds use are mapped in their memory. These places become so routine that many trees and shrubs have their bark blistered off from the constant wear and tear of landing and perching.

Flight and its endurance in lift is used also by many vascular and nonvascular plants of the plant kingdom. A most common sight of childhood is the parachute wind takeoff of dandelion seeds from the ray disk platform of the flower head. Few children have not blown these parachutes to the four winds with nursery rhymes. But there are still examples of lift finely attuned to seed design. This is seen in the samara seeds of the *Aceraceae* or maple family. These samaras have an angle of separation between the twin samaras and this develops the rotational ability of the samara to spin and to take flight. Another sneaky system is used by the mosses. The calyptra is the fruit of fertilization, a capsule with a pointed cap. It has a circle of specialized cells that are exquisitely sensitive to air moisture. When conditions are right, the calyptra acts like a giant catapult and spores get fired by torque over incredible distances.

Many birds have evolved different likes and dislikes of lift, from the golden eagle to the hummingbird. The elusive bluebird catches the first rays of light of the early morning from its easterly-facing cavity home. This bird likes to have a nest that is not higher than six feet (two meters) above the ground. And the sloe-eyed vireo disappears into the fortified solitude of the white ash, *Fraxinus americana*, profile. The white weavings of its nest almost at head height reflect the dappling of the pinnation of the leaves to perfection. The nest and its contents are invisible.

The branches of the trees of the global forests provide something of vital importance for birds. As the branches reach in their own particular way for solar solace, they, too, make way for solar exposure. All birds need this exposure to the sun. The sun performs a molecular miracle for the birds. In an ordinary average spring day the flight feathers of a bird become greasy. Dissolved in this grease is a precursor form of vitamin D, a fat-soluble and essential vitamin. The bird, any bird, perches in the sun to sunbathe. The energy in the sun's photons breaks a double bond in the precursor vitamin D, changing it into a full and functioning form of vitamin D itself. The bird then preens itself. In cleaning its feathers the bird ingests its quota of vitamin D. This vital ingredient for egg laying sets down the best foundation for reproduction for a bird, a successful hatch.

In the lazy lope of one of our most beautiful pollinators, the butterfly, lift is supplied by something else. In recent years entomologists using radar began to work with meteorologists, and discovered that high-altitude air currents provide lift for butterfly migrants. They were described as "a river of butterflies," out of Panama. The butterflies make good use of the seasonal patterns of weather change common on the continental face. It was also found that more than the monarch migrates—these other migrants include the sulfur and the *Vanessa* butterflies. The latter are also more commonly known as the painted lady and red admiral butterflies.

The monarch migrant, prior to transit from its Mexican arboreal staging ground, muscles up into 125 percent of its dry weight as fat tissue. And this, too, is vital for egg laying for this species. Sex is sworn off for the trip. The monarch fires up its extraordinary navigational system, that of solar polarization. The compound eyes of a butterfly can read the mixed messages of polarized light and then use it as a navigational tool. In the spring the sun sends out flares. These solar flares polarize or twist the sun's ordinary photons. This changed light pattern converts visible light into an invisible form but one that can be perceived by butterflies. And butterflies see these images made by polarized light and use them as a focus for orientation in navigation.

The world of polarized light forms a different landscape for a butterfly. All plants and especially trees absorb and reflect polarized light. The cell walls of cellulose and hemicellulose are set in a distinct pattern to do this. As the sun and its flares move up on the horizon into a northern summer, the landscape resonates with this light which is visible to the butterfly and for which the human eye needs polarizing lenses.

The male butterflies race to get their first place in the sun while the females fret and float to conserve their fat for egg laying. The waiting males get the first pick in their dance of life. Some, the more early azures, pick an apical meristem of a nut tree, while others go farther afield. But all use the same homing device for their own particular pattern of egg laying, that polarized light from the sun. And lift, the seasonal drift of airmass, is there again to move the winged beauty as it is for the scramjet, as it is for the bird . . . the envy of man.

All Trees Hold the Dream of the Forest

DREAM WORLD

Dreams are the currency of sleep. They arise from the great pool of wisdom that rests in the subconscious mind. Dreams have held a special meaning to almost every culture tied to the global forest.

The biochemistry of the dream pivots on a chemical called melatonin. It is an aromatic hydrocarbon. It is produced in a tidal pattern to comply with the cycle of the movement of the sun and moon in the sky. This rhythm of night and day activates and suppresses melatonin, which produces sleep from which the dream arises.

All trees hold the dream of the forest, too. Trees produce a hormone similar to melatonin called auxin. It, too, is an aromatic hydrocarbon. It is produced in response to the changes in sunlight of the seasons. Darkness plays a role, also, especially underground in the roots. These photoperiods, in rotation, balance sleep or respiration in the tree so the dream can arise.

Medicine men and medicine women of the forest can catch these dreams. They can also focus into them and amplify their meaning. The meaning that comes out of a dream has great importance. Often the dream is personal to the dreamer's life. At other times the dream is vital to the culture at large. This kind of dream is usually the product of a dream quest pursued at a special time by a medicine man or woman.

In many parts of North America a dream quest involved the ritual of the use of a dream stone inside a dream circle. A dream stone is a very large, flattish stone that can accommodate the full length of the human body in a reclining position. The stone is usually granite with two hollow areas, one for the head and the other for the hips. These stones are always bare. They never carry a lichen encrustation because of centuries of use. The patina of the oil from the skin prevents any plant growth. The central dream stone is surrounded by a wide circle of smaller stones placed at a given distance radiating from the larger one. Each of the smaller stones in their circle never touches. They are always rounded and smooth. The dream stone assemblies are always in a forest setting because of silence.

There are other means of fine-focusing a dream for the common good. These dreams are dreams of prophecy. They are capable of looking into the long fetch of the future. They are made by honored medicine men and sometimes shamans of various cultures throughout the global forest.

These dreams are initiated by mind-altering drugs that are used in sacred doses and in ritualistic ways or through fasting, dance, and meditation. At present one such forest biochemical is still in use for the production of a dream forecast for the future. It is a special aerosol that is liberated into a fine smoke by the slow smudge burning of a forest fungus called the tinder fungus, *Inonotus obliquus*, also known as *Poria obliqua*. This strange, blackened mass of mature, protruding, fruiting bodies is burned. The smoke itself carries the special mind-altering chemistry. This is carefully inhaled. The effects travel in the mind to form the dream. This very rare, sacred plant also occurs in the tropical forests of Sri Lanka.

Fasting and death produce potent dreams. Often these are considered to be sacred. There is one sacred dream making its way in the world of today. It comes from an elder of the Hopi people. It is a vision quest of death. Before he was due to die, the elder asked that the people of all nations pay attention to the trees of the forest. One such tree entered his dream. It held a circle of light in its trunk. The light came from the power of the sun. This was transferred to all the creatures of the earth as a life force. This dream is a warning to all people to respect nature, beginning with the trees of the forest, because the web of all life as we understand it depends on the tree.

In times of sickness and disease the plants and trees of the world spoke to the medicine men and women in the form of the dream. This has happened all over the world, from the Americas into Russia, the Balkan states, and on toward India and Asia. The plants and trees revealed the medicines they held.

Sometimes the cures were new. Other times they were mixtures of known and unknown plants and trees. The medicine man or woman was sent to search out the new plant to add a greater potency for a cure. In all cases they were told from what part of the tree they could take scrapings. Those from the southern exposure of the trunk carry the greatest medicine. The samples from the northern exposure have the least. This is reversed in the southern hemisphere. The roots were used, the flowers, the leaf bud while still enclosed in the bud scales, and sometimes the leaf itself in a mature state.

This medicinal knowledge became the oral academy of the medicine men and women. There are eight major medical pharmacopoeias in North America. There are more in the tropics and an even greater amount in the cold of the boreal north. These sources of natural medicines were protected by a harvesting rule of picking so that there would be sufficient left for the seventh generation into the future.

These medicines became the foundation knowledge of most modern medicine. It is from the wild that the unexpected knocks on the door of opportunity for health . . . from the wilds of nature and the wilderness of our dreams.

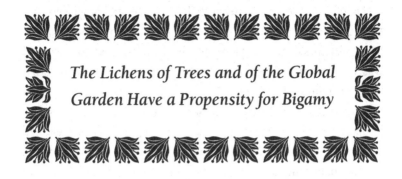

The Lichens of Trees and of the Global Garden Have a Propensity for Bigamy

THE MARRIAGE OF LICHENS

Lichens are like time and the tide; they will wait for no man. Lichens are found growing on trees, as swags on branches. They are found on rock surfaces on intertidal marine areas and on into the vast landscapes of the north. They grow on soils and on fence-rows. There are even lichens that grow on lichens. There are others, a little fussier, that inhabit skulls.

The story of the lichen is a story of scientific mystery. Although a lichen is composed of two separate individual and unrelated plant species bound in a marriage of convenience, we do not know where to place them on the evolutionary tree. The marriage of a lichen is between an algae and a fungus. Between them they make a home called a thallus. But there is a sinister aspect to this marriage, because the union does not enjoy the usual conjugal happiness. Rather, it looks a bit like slavery with the algae partner acting as the slave and the fungus behaving as the master. In another twist, it looks as if the fungus discovered agriculture and the seaweed, the algae spouse, is the crop.

There are fourteen thousand lichens in the global garden that spread from the tropics almost to the poles. They will crawl and live as hermits in arctic deserts. Once upon a time when the world was young they enjoyed a remarkable glory. They shone through the virgin forests like colored gems. They set sail from the rugged trunks of trees catching the winds. And out of these winds the haustoria of the lichen drank the morning dew. They glistened with moisture and measured it inside their prettily colored homes. On the most venerable of trees and from the out-thrown branches the lichens extended their limbs downward in sylvan sleeves. The ancient, virgin forests wore the lichens on their backs like majestic jackets.

There is not much known about the fungi that form the perimeter walls of the thallus structures. They are from the most advanced species in the fungus kingdom. They are the sac or ascus bearers whose whole clan are the clever ascomycetes. These reproduce by heavy and dense ascospores. These unknown fungi capture simple plant creatures from the ancient seas and hold them captive in an extraordinary coupling of the lichen world.

The sea spouses are the blue-green algae that are sometimes called the cyanobacteria because they, like all mammals, hold glycogen in their bellies. Sometimes other algae are chosen. These are the unicellular ones of the upper ocean. They are found on land, too, in damp and dark places. Occasionally a touch of bigamy takes place and two species of algae are found in one bed.

The algae, though they are the underdog, define the shape of the household or thallus housing. So in a way they define the species of lichen itself. The algae work in the green kitchen of photosynthesis. The chlorophyll is attached to membranes called thylakoid membranes. These structures are very primitive. They are not like the tight superior boxes of the chloroplasts of the vascular plant world that are governed independently by their own smidgen of nuclear DNA.

The chlorophyll of the algae does what all other chlorophyll does. It sups on carbon dioxide and water and produces food. The food is the usual sweet sugars and alcohol complexes of plants. This food is needed by the fungus partner, who cannot make food from photosynthesis because it lacks the machinery to do it. So the fungal hyphae move close to the algae cell and by fair means or foul manage to make this haustorial cell become porous to sugars. From the unicellular algae a sorbitol sugar is switched by osmotic pumping. From the blue-green algae a glucose solution is obtained. Satisfied, the fungus then gets busy and chemically cranks these sugars into a store called mannitol. It is on mannitol sugar that the fungus actually feeds and thrives.

The algae component does something else that is important in its life's work. It fixes nitrogen. It produces an enzyme that is capable of taking nitrogen out of the air and making nitrates and nitrites. This is a passageway to protein manufacture on which all life lives. The algae must, of course, share some of that protein with the fungus. But on the larger stage, the algae shares it with the world in its life and death cycle. This is as important on land as it is in the oceans. Again it is a numbers game; the larger the number the greater the effect. This is most true of the boreal forest region of the world where lichens are so very dominant. They sequester carbon dioxide out of the atmosphere and regain a nitrogen fixation in a local nitrogen cycle, which has global importance to northern feeding and breeding cycles of life. This balance of biochemistry is as yet poorly understood and forges an important reason why the boreal world should be protected.

All species of lichens reflect a remarkable range of chemistry. Many organic chemicals are unique to them alone. This array of molecular activity is born out of their marginal life and reflects their ability to survive in the most desolate and hostile environments of the world.

More than six hundred organic chemicals produced in lichens have been discovered. They all have a function within the home or thallus. Some like the xanthones and the pulvinic acids are pigments and are found on the sunny outside of the thallus. They act as photoamplifiers for the chlorophyll membranes, increasing efficiency. Others are foul and deter browsing ungulates.

Many more have an antibiotic function, which inhibits foreign fungal spores or even stray seeds from germinating. A host of chemicals keeps the thallus dry and maintains a free flow of air throughout the structure, which is built with an extraordinary molecular water-shedding ability. And one other unusual lichen, *Umbilicaria esculenta*, rock tripe, holds the HIV virus at bay.

The Forest Holds Many Secrets of Sex

GREEN SEX AND THE AFFAIRS OF THE HEART

This is by way of a warning. The sexual life of a plant is not for the faint of heart. Lessons of licentiousness are not just found in the red light district, they are alive and well in the plant kingdom.

Then there are the tales of the heart. These abound. Some plants make astonishingly good mothers while others are mean and miserable.

A class of forest mosses cage their children spores in a sac called a calyptra. They remain in prison until the weather is ripe for rain. As the dry days end, a set of ceiling bars go into action. These are called the peristome teeth. They are made of a carbon-based material unique to science. They retain an enormous torque action which increases in strength during the dry days of summer. When this changes to rain, the bank of teeth opens, becoming hydrated, and a powerful ejection system is set in place. The spores are dispersed far and wide into the jaws of a hungry world, while the moss mother plant relaxes back in relief.

There are the heroic mother plants of the hellebore world in early spring. The plant will burn off its excess energy in calories to melt the snow around its growing buds. When they reach the warmer air above the ice and snow, the mother plant pushes forth further heat to get the flower in all of its breathless beauty to be pollinated by early bees.

The southern American basswood likes its taste of flesh. *Tilia tomentosa* sneaks a hypnotic potion into its floral nectaries because it is hungry for nitrogen. The pollinating bees take a hit. They slowly die in the cool shade, their bodies supplying just enough nitrogen for the trees to make maternal protein for another season of sex.

There is a sexual tale coming from southern China that really puts the icing on the cake. This is a forest home of great unevenness where cliffs and rocks intermingle with the trade of trees. It has been said that in these tropical cauldrons the steam rises and it is not just from the humidity in these overheated places. The scientists involved are still scratching their heads and putting the matter to boil in their pensive brains.

It would appear that in the tropical forests of southern China among the playful platycarya, a real zinger of a family exists. This touchy-feely family goes by the name of *Zingiberaceae* or the ginger family. This redhead is related to the modest and puritanical ginger, *Asarum canadense*, and *A. virginicum* of the North American woods. These lace-faced maidens have earned their living by body part donation to medicine on the altar of North American health.

The Chinese diva in question has the glass-breaking name of *Caulokaempferia coenobialis*. So far a common name has not been attached to this plant, its gown or its entrails. This strap-leaved lassie embraces life near the forest floor. But her big penchant is for large rock faces from which she hangs by hairy tendrils. She sings her sexual song horizontally from a vertical rock position. Her embrace is absolute. Her brassy character is well established by her ability to withstand a constant drip of moisture into and about her private parts as they develop into maturity.

It comes as no surprise that in this place of rocks and trees in the middle of nowhere her sexual partners are scarce. The circumstance of constant moisture does not improve matters either. But the drive carries on nonetheless in the rat race of chloroplastic cuisine around her. So when the unhappy day arrives and she is pining for a partner, she becomes innovative. She takes the mother of all invention by the hand and does something that no other plant has ever dared to do. She initiates the entire process of sexual satisfaction herself.

In a determined fashion when her ova are just ripe, *Caulokaempferia coenobialis* releases an oil slick from her female parts, the gynoecium. She does this when she is in a state of relaxation and the afternoon's sun has intensified the heat of the day. She sits simmering in the heat of the sun, fully exposed for the situation to come. The oil slick increases and spreads with the aid of this solar sympathy. The slick gropes across her petals and soon picks up her waiting pollen, whose male stance is that of ripeness and waiting from her other private parts. The heat is turned on by the afternoon. The oil relaxes with an increase in temperature to produce a decrease in viscosity and an increase in flow rate.

The inevitable is about to happen. It always generates paternal palpitations. But there is one further uphill battle because the diva herself is hanging vertically and nothing flows uphill. Not until now have Mr. Newton's laws been put to such a rigorous trial.

The pollen-bearing oil increases in girth in a concentric manner. This happens with all the attractive forces of physics up the diva's sleeve coupled with surface tension. This mobilizes the face of the oil in its roil toward the mating game.

Suddenly the ante is up. The oil has reached the ovum. The dividing pollen pulls up its sleeves to produce a pollen tube, and in no time at all the nuclear load is moving. The chalaza has received its fertility farewell and a new life is on its way. The diva has outsmarted them all in her talent to ensure sexual penetration. Indeed, there is none in her class that could outsmart one law of physics, that of gravity, by using another law, that of quantum mechanics, in such a molecular manner to achieve her ends.

A sheer analysis of such an extraordinary sexual oil would surely indicate that its flow rate is perfectly matched to movement on the petal surface of the flower. It would flow in such a way to ensure outward momentum which was equal to the forces of gravity.

The warning is over; you can open your eyes now. The diva is only related to North American plants by way of chemistry, medicinal biochemicals. Come to think of it, that is another story too.

*Forests Form a Basic Barrier to the
Pillage of Pollution*

DIRTY LAUNDRY

There is a new violence in the world. This violence stands apart from all the other familiar forms. This one is silent and it shows no mercy to the young and the old. It is in the air we breathe. The air is no longer clean.

The new violence is measured in microns, a size smaller than a pollen grain. The violence is particulate pollution. This form of pollution is composed of tiny fragments of matter that will become airborne. Anything can become airborne if it is small enough or light enough or has the right kind of aerodynamic form to fly. These tiny particles are now finding a new name depending on their diameter, expressed as particle microns, or PM for short.

Size matters. Any pollution particles that measure PM 2.5 or less are lethal to the human body. They are also lethal to much of the rest of the animal kingdom. They are generated by industry, incineration, fossil fuel burning, traffic, and global drought patterns mixed with wind and by warfare, explosions, and volcanic eruptions. Airborne pollen can add to their vengeance.

The particles accumulate and concentrate in the air of cities and urban landscapes. The particles that accumulate together in mass movement are known as smog. Smog days are occurring earlier in the year—mid-March in North America—and more frequently worldwide.

Particles of 2.5 microns or less irritate the lungs and damage the entire circulatory system. They can reduce birth weight and damage the brain. Metal-carrying particles accelerate asthmatic constriction. These particles go into the deeper passages of the lungs—the tiny bronchioles, where the body begins its oxygen extraction from air. These bronchioles are paper thin and delicate. They have to be to do their work. In the presence of the pollution particles, the process of extraction is hampered and the lung tissue gets irritated. The lungs can produce free radicals to fix matters, but this in turn causes scarring of the lung tissue, a condition called fibrosis. The natural act of breathing becomes more difficult with fibrosis.

The story for the heart is similar. The arterioles get clogged with the particulate pollution of 2.5 microns or less. These smallest of arteries deliver oxygen-rich blood to the local tissues. Such paper-thin walled arteries must relax a little further to complete their oxygen delivery. They do this by means of a dissolved gas called nitrogen oxide. But, in the presence of particulate pollution of 2.5 microns or less, the nitrogen oxide doesn't work, again because of irritation, and the arterioles cannot relax and deliver oxygen. This causes damage to the local healthy tissue. The arteries also have a regulatory protein called endothelin, which controls blood pressure locally. The particulate pollution causes this protein to increase. In normal tissue this is not a problem, but in atherosclerosis where the arterial wall has changed, this causes heart attacks and strokes. Nasal passages, too, break down, leaving areas of the brain open for contamination by brain plaques.

Hitchhikers travel on these particles like magic carpets. The hitchhikers can be hydrocarbons of various kinds, metals or pesticides. Other toxins like dioxin and furans can travel and work together, making their visit more deadly. Metals like spent plutonium from modern weapons with its almost eternal half-life can make the trip as can various other jockeys like vanadium, titanium oxide, and lead.

Trees and forests hold the answer to particulate pollution in a way that is surprising. Many trees have leaves that differ from one species to another. This diversity is found in the leaf's anatomy. Some leaves have a waxy cuticle on their upper surface. These leaves repel water and attract particles that are water insoluble. The undersurface of the leaf is downy. This down is composed of thousands of fine hairs, all only a few microns in size. These hairs are multiplied in the full canopy into billions of fine hairs.

Leaf hairs themselves have further characteristics. Sometimes they are fat. Other times they are tall and skinny. Quite often they arise from a larger pedicel and decline in size to a pin-shaped tip. On some occasions these hairs can be floppy and on others they can be stiff and erect. Their spacing on the leaf surface changes too. The more downy surfaces will have a greater distribution of hairs and the less downy leaf will have fewer hairs. In addition, the microscopic surface of these leaves is full of ridges and folds. These form steep valleys with the midrib and other subsidiary leaf veins. The pattern of these veins might be in the form of a net, or they might run parallel to one another as in monocots.

This microscopic world of the leaf within the tree canopy acts like a fine-toothed comb for the air. The particulate pollution of the air becomes caught mechanically like dandruff in this microscopic world of hairs. Sometimes the particles, which hold a charge, can get grounded on the tree. This depends on weather conditions, and on electrostatic forces of attraction generated by the tree's leaves. The trees and the forest act as a sweeping brush or giant comb. The leaf hairs numbering in billions clean the air of these tiny particles. These particles get swept down the trunk by rain and are detoxified by the hungry microbiota of the living soil.

A healthy tree with a wide canopy around a house will significantly reduce particulate pollution. Urban forests and forested areas in our neighborhoods, parks, and cities have the same effect on the urban environment. Global forests do this on a planetary scale. They form a living wall for health and a basic barrier to the pillage of pollution.

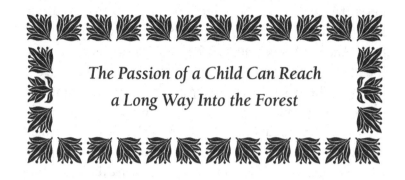

PASSION AT PLAY

The child stood looking up into the tree. From where she stood the tree was so tall, like some church spire probing the sky. She tried to get a glimpse of the tip of the tree. It made her five-year-old legs stagger and buckle. She stepped backward while still staring into the crisp blue sky, sheered with the curtain of river mist. And still there was one more fruit to fall. Just one. She would wait until that last tiny apple fell and the tree, her tree, not her favorite tree, her second favorite, would have all of its apples on the ground in a red-yellow carpet around her feet. She would then stand in an apple bed of freshly fallen fruit.

She had been told something delicious a week before. Her tree, her apple tree, was not an apple tree at all. It was a *Crataegus*, an American hawthorn. An American living in her garden. It had come all the way from America to live in her garden. That American had chosen to live next to her very favorite, her all-time favorite, tree. In her own garden. She savored the word, *Crataegus*, repeating it under her breath until she knew that she would not

155

forget the name. It was her secret word. It was a tree word. She waited patiently for that last apple to fall. She waited all afternoon. And just before teatime it fell with a plonk. A dead kind of sound of something within its own body, that gave a little bit of itself with the long fall from the top of the tree. She knew this last one, this final apple, would be bruised. It would show a circle of brown flesh that would spread outward with time into the white flesh of the apple. The bruise would consume the whole apple in time. She decided that she wouldn't give the apple that time. That final apple would be the apple for her. That would be the apple she would eat.

As she ate the last apple, she surveyed her kingdom. She knew that the kingdom surveyed her, too. She was welcome here. She was part of this garden. She owned this garden because she was a witness to it. Because she was a witness she understood it like nobody else. That was why she laid claim to its ownership in her heart. Because, simply, it was hers.

As she finished the apple she held the core in her hand. It was like an ordinary apple, a bit smaller, but an apple nonetheless. She had eaten the bruised part, too. The bite had fragmented on her tongue. She knew she was right to eat it before it went brown and sour. It saved the apple from decay. She had saved one apple from that tree and that made her happy. She did this act of kindness for the tree. Next she turned to look at her favorite tree. She smiled at it. The tree was five skips away. It was the most favorite tree in the entire world. There was no better tree than this tree ever to be found anywhere. You could look in books, you could examine other gardens. No. She was satisfied that she was right. This tree was the best. The very best. It was the king of all trees. It was the king of the kingdom. Her kingdom. Her garden.

She had known the name of this tree, her favorite, for as long as she could remember. And that was a very long time. She had been asked to pick the leaves for her semolina pudding. She didn't like to pick the leaves because she knew that she hurt the tree by taking the leaves from it. She imagined that the trees of her garden did not like to share their leaves. It was mean to take even one leaf. But she took the leaves willingly because she liked the taste of the leaves in her pudding. The taste was good. She never ate very much. That was why she was so thin. But she did like the taste of these leaves and the flavor they left in her mouth. Without these leaves, she thought that she wouldn't ever feel like eating at all. Ever really.

She understood her favorite tree. It had a silky trunk. It was gray and smooth like the skin of an elephant at the zoo. When touched, the gray had a feeling of coolness and an odor of freshness that came off the trunk itself. She knew. She smelled the trunk. Nobody else had done this, of that she was sure. The tree had tiny yellow-green flowers in the spring. There was never a flower that was alone. The flowers were in bunches. She had never picked these flowers. She had never wanted to. They were part of the tree. Anyway she had lots of bluebells to pick and that was enough for her.

In the autumn, when it became dark and late and scary, her tree held a secret. It was a black secret. It was seeds. They were round and solid. They had a coat around them like a black olive. And they had a round seed, too, at the center. Once she stepped, by mistake, on one of these seeds. It squashed flat. But it produced a very vivid smell. It was a kind of smell that she couldn't place. She didn't like it. So she made sure that she never did it again. And anyway one spring, too, she found one of these seeds sitting on the gravel and it had a little white root and two small green leaves. The skin was curled backward like a black cap. Gently she carried the seedling in both hands to a very dark place in the ground where it would be safe. She hoped it would be safe there forever.

She looked at her favorite tree again. She was satisfied about it. She knew all about this tree, even its name and even its family. She repeated to herself her mantra of knowledge, *"Lauraceae."* A sweet bay tree. A *Laurus nobilis.* "You are the *baccae lauri*, the noble berry tree. And . . . one day I will learn about you." The passions of a child reach deeply into the forest.

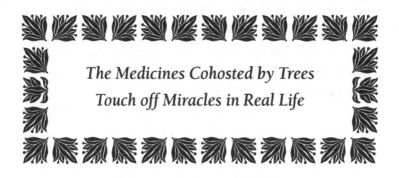

*The Medicines Cohosted by Trees
Touch off Miracles in Real Life*

THE MEDICINE WALK

It has been estimated that almost half of the medicine used by the world's human population comes from the vast array of plants of the plant kingdom. The other half are molecular mimics of what has already been here, for design in chemistry is regulated by the imagination, too.

Outside of the human family, every functioning species on the planet requires the salve of medicine from time to time, like the bird that uses dry soil as a dust bath for feather lice or the canine that ingests snippets of *Agropyron repens*, couch grass, for the antibiotic it contains. Or the beewolf wasp, *Philanthus triangulum,* which uses a white material marker for orientation and for the production of an antibiotic that keeps a unique strain of streptomyces at bay in its brood chambers. The beneficial use of plants by mammals for medicinal purposes has been mostly ignored. Both plants and animals are based on similar biochemical models. The metabolic pathways for the functioning of plants are similar to those of mammalian systems. There is probably a synchronicity between the two that has never been explored.

There is also, in all probability, an interdependence that is being missed because of its obvious nature. The plant species that live for the longest time quite often have the greatest treasury of medicine at their fingertips. These plants are the trees of the global garden. Out of this treasure trove some chemicals have had an astounding effect and changed the course of history in the global garden.

One mind-altering chemical called ergotamine was active on both sides of the Atlantic at the same time. This is now used in the treatment of migraines. In North America, for the peoples of the boreal forest, ergosterol was a sacred drug that could lead the mind into divinations for the future. It was manufactured by an advanced order of fungi. One of these fungi, the tinder fungus, *Inonotus obliquus*, also known as *Poria obliqua*, colonizes the northern species of the *Betulaceae*, birch, and the *Aceraceae*, maple families. A chief or an elder used the dark hyphal bolus as incense. It was slowly burned as a smudge. This created aerosols that contained ergosterol derivatives. The smoke altered the brain functioning of the elder or chief. The circulation of the arteries of the brain was changed and a pattern of the future was seen. This ritual information was then used for the general welfare of the people as a whole.

In Europe as far back as the sixth century another ergotamine-producing fungus was busy. This species cohosted on a cereal grass called rye, from which flour was and still is commonly milled. The fungus that was the most active was *Claviceps purpurea* or ergot. It occasionally had help from a cousin called *Cordyceps*, which, too, was called ergot. These fungi, when they were fed up with the forest's arena, took aim at the juicy starch of the developing ovary in the grains of rye as they moved toward maturity. In the infected grain a somewhat dark and deadly body was produced. It was the resting stage of either fungus in its cohost travels. The body itself became a sclerotium. This sclerotium, when opened, exposed a black mass similar to the North American tinder fungus.

Fields of rye, *Secale cereale*, were cut and harvested. They were milled into rye flour and baked into the usual dark rye loaf of bread that was placed for cutting on the kitchen table. Throughout Europe the people who ate this infected bread developed a strange form of gait. They looked for all intents and purposes like people who were continually hopping over a hot fire. So the dance they produced was called St. Vitus' dance. This dancing festival lasted until the beginning of the nineteenth century, when the weather got dryer and the culprit flour was recognized and then blamed.

For over a millennium, ergotamine was also the interesting subject of miracles. The people of Europe who had a particularly tough form of the local hop went on a pilgrimage. They found that when they got as far as a hovel in which a French monk had set up house, they became remarkably better. The hop halted. So a shrine was erected. At this shrine all hops halted. This was a miracle, indeed. The monk's name was Anthony and he became St. Anthony. All who had their hop halted recognized St. Anthony in their prayers of thankfulness. In due course St. Anthony took ownership of the hop as well as everything else that was truly miraculous. The whisper went around that the hop was also called St. Anthony's fire. He had the lure at his own personal shrine.

But the cure itself came in a different mode. As the pilgrims hopped as far as St. Anthony's shrine, the local weather was dryer. The sclerotia of *Claviceps* and *Cordyceps* need dark damp weather to go into their resting state, so the flour was free of contamination and so was the bread which was eaten in very generous amounts on the pilgrims' journey. The body excreted the ergotamine in colorful urine. And the hop halted at the gate of the shrine.

The United States of America had its taste of miracles too, in 1692. But these events took another turn, from hopping to swinging. The rye and its cohost fungi traveled to Salem with the pilgrims. The weather turned damp and humid. The fungus produced high-quality sclerotia and the women met their fate at the witch trials. A vision of the future is not always a good thing. . . .

In Every Ending There Is a
New Beginning

THE FOREST AND THE FIRE-KEEPER

Unlike books, words are old. So are the oral traditions of global cultures. A collective memory was kept alive and passed on in the oral traditions of storytelling, songs, poetry, and family bloodlines. Occasionally within a culture a mystic would arise who could see a clear vision of the future from the patterns of the past. These mystics were prophets. They had the gift of prophecy and their words were remembered.

The oral tradition of any culture is its essence. It skims off wisdom and keeps these nuggets close from bosom to bosom down through the generations. This wisdom is collectively gathered and is committed to memory for use. Those who gather this wisdom have many names in many cultures. For the First Nations, the aboriginal peoples of North America, they are called the "Fire-keepers."

The fire-keeper is the keeper of legends. The fire-keeper is also charged with the memory of prophecy. Such items of future history are recorded according to a dateline of some event outside of the normal measurement of time. This clocks the event into the future in a timeline that nobody can dispute.

The forecasts of prophecy are fairly common phenomena that are often adopted universally and become embedded in the public's consciousness of the time. These, too, are remembered in common as an anchor for the prophecy itself.

At present there is a prophecy for the direct future of the times in which we all live. This prophecy concerns nature itself. Nature or Mother Nature, which is sometimes described as Gaia, is composed of a complex web of life in which all things live in an interdependent manner. The small is equal to the large in this network of life. And the small is codependent on the large for its life force. Every cog is placed in every wheel for a reason. There is a balance there too, a little play in this giant system so that it all works together hand in hand.

There is a timeline for this prophecy, too. The prophecy will happen around the time of the great dying of the North American maples, *Acer saccharum*, the sugar maples. These maples are the great feeding trees of the eastern seaboard of America. These trees will begin to decline from the tip. At first the tops of the trees will wither and die. Then the disease will spread downward through the trees until they lose all of their leaves. This dying is the beginning of the timeline of the destruction of nature.

The rape of nature has then begun. Other trees will succumb to various infestations. The loss of the forests will foreshadow a period of devastation. People will not realize what they have done, but they will continue in their path of demolition. From the peoples of today will arise another new generation of children. These children will be different from all those who came before. These children will have many gifts. They will be able to do extraordinary things.

Primarily these children will have the gift of telepathy. They will be able to communicate with one another across the globe, even though they do not know one another. Their recognition factor will be youth itself. These children too will have the gift of the dream. In the dream they will have clarity of vision. From this dream they will understand what has happened to nature. They will understand it and comprehend what their parents have done. Many of these children, too, will have the gift of prophecy. This will frighten them in the beginning until they gain an understanding.

Then the children of this generation will want to help the planet and nature in a collective way. They will hold hands across the planet in their minds. They will alter their parents' ways. They will encourage one another. In this circle of life the children will save their parents through a dream and through a prophecy. In saving their parents they will save the planet.

This is an old legend, told before the advent of the computer or the Internet. It was told before the advent of radio, television, mass media—even before electricity. It was told at a time when the sugar maples were healthy and producing copious quantities of sap for maple sugar.

Even as the words of the legend of the fire-keeper come together there is a truth to them. The media is filled with stories of nature's abuse. Those who should protect nature calmly put whole forests like the boreal forest on the chopping block without a moment's hesitation. Those who want more oil are busy with their killer sonar techniques in the krill-rich waters of the Tatarskiy Proliv

(Strait of Tatar) between Sakhalin Island and mainland Russia, the birthing ground of the bowhead and the feeding grounds of the great whales. There seems no end to greed and no beginning to sustainable management of the planetary resource basis.

But the children exist. They have been taught a better mode of planetary management. The consumerism in their lives bores holes of unbearable solitude. They are already reaching for something else, something elusive, something that is color-blind to race. It is called dignity, the dignity of life, all life.

ACKNOWLEDGMENTS

There are people to thank and to be grateful that they are alive. One of these is my husband, Christian H. Kroeger. We work as a team. I write and he does his exquisite photography. He edits my work with a gentle hand, so that my voice is heard. Few can do that. I also want to warmly thank Mrs. Nancy Wortman for her skill in production and her cheerful presence. I thank Mr. Lynn Wortman, her husband, too. There are two more people in New York. The soft-voiced Paul Slovak, my editor, who heard my dream. And of course, Stuart Bernstein, my agent, the best of the best. Without these people around me, I doubt that this work would see the light of day, for closeness amplifies the work of the mind to bring forth the sparkle of creation.

REFERENCES

Arnold, Elizabeth, and Keith Clay. "Sweet Lurkers, Cryptic Fungi Protect Chocolate-Tree Leaves." *Science News*, December 2003 (vol. 164, no. 24), 374.

Baldwin, I. T. "Chemical SOS Not Just for Farm Lab Plants." *Science News*, March 2001, 166.

Beresford-Kroeger, Diana. *Arboretum America: A Philosophy of the Forest*. Ann Arbor: University of Michigan Press, 2003.

———. *Arboretum Borealis: A Lifeline of the Planet*. Planned publication, fall 2010.

———. *Bioplanning a North Temperate Garden*. Kingston, Ont.: Quarry Press, 1999.

———. "The Black Willow." *Grand Traverse Band News*, July 2008, 21.

———. "Exceptional Shade Trees." *Nature Canada*, spring 2005, 33.

———. *A Garden for Life: The Natural Approach to Designing, Planting, and Maintaining a North Temperate Garden*. Ann Arbor: University of Michigan Press, 2004.

———. "King of the Forest." *Nature Canada*, spring 2000, 16–19.

———. "The Oak." *Nature Canada*, spring 2005, 29.

———. "Preserving the Butternut Tree." *Eco Farm and Garden,* spring 2003, 44–47.

———. "A Summer Beauty." *Nature Canada*, winter 1999, 18–19.

Boon, Heather, and Michael Smith. *The Botanical Pharmacy*. Kingston, Ont.: Quarry Press, 1999.

Borror, Donald J., and Richard E. White. *A Field Guide to the Insects of America North of Mexico*. Boston: Houghton Mifflin Co., 1970.

Brodo, Irwin M., Sylvia Duran Sharnoff, and Stephen Sharnoff. *Lichens of North America*. New Haven: Yale University Press, 2001.

Budavari, S. *The Merck Index: An Encyclopedia of Chemicals, Drugs and Biologicals.* 11th ed. Rahway, N.J.: Merck, 1989.

Casselman, Bill. *Canadian Garden Words.* Toronto: Little, Brown, 1997.

Chenoweth, Bob. "The History, Use and Unrealized Potential of a Unique American Renewable Nature Resource." *Northern Nut Growers Association Annual Report* 86 (1995): 18–20.

Chrisholm, Sallie, and Nicholas H. Mann. "Probing Ocean Depths, Photosynthetic Bacteria Bare Their DNA." *Science News,* August 2003 (vol. 164, no. 7), 100–101.

Clausen, Ruth Rogers, and Nicholas H. Ekstrom. *Perennials for American Gardens.* New York: Random House, Inc., 1989.

Cody, J. *Ferns of the Ottawa District.* Ottawa: Canada Department of Agriculture, 1956.

Collingwood, G. H., and Warren D. Bush. *Knowing Your Trees.* Washington, D.C.: American Forestry Association, 1974.

Cormack, R. G. H. *Wild Flowers of Alberta.* Edmonton: Queen's Printers, 1967.

Cox, Paul Alan, Susan Murch, and Sandra Banack. "Plants, Bats Magnify Neurotoxin in Guam." *Science News,* December 2003 (vol. 164, no. 23), 366.

Cox, Paul Alan, and Oliver Sacks. "Troubling Treat: Guam Mystery Disease from Bat Entrée." *Science News,* May 2003 (vol. 163, no. 20), 310.

Dahm, Werner J. A. "Soaring at Hyperspeed, Long Sought Technology Finally Propels a Plane." *Science News,* April 2004 (vol. 165, no. 14), 213–14.

Davies, Karl M., Jr. "Some Ecological Aspects of Northeastern American Indian Agroforestry Practises." *Northern Nut Growers' Association Annual Report* 85 (1994): 25–39.

Densmore, F. *Indian Use of Wild Plants for Crafts, Food, Medicine and Charms.* Ohsweken, Ontario, Canada: Iroqrafts, 1993.

Edwards, Martin, and Anthony J. Richardson. "Early Shift, North Sea Plankton and Fish Move out of Sync." *Science News,* August 2004 (vol. 166, no. 8), 117–18.

Fell, Barry. *Bronze Age America.* Toronto: Little, Brown, 1982.

Flint, Harrison L. *Landscape Plants for Eastern North America.* New York: John Wiley and Sons, 1983.

Fox, Katsitsionni, and Margaret George. *Traditional Medicines.* Cornwall, Ontario, Canada: Mohawk Council of Akwesasne, 1998.

Frazer, James G. *The Golden Bough*. New York: Avenel Books, 1981.

Granke, L. J. "Genetic Resources of Carya in Vietnam and China." *Northern Nut Growers' Association Annual Report* 82 (1991): 80–87.

Hamilton, J. W. "Arsenic Pollution Disrupts Hormones." *Science News,* March 2001 (vol. 159, no. 11), 164.

Hashimoto, Kimiko, and Yoko Saikana. "Red Sweat: Hippo Skin Oozes Antibiotic Sun Screen." *Science News,* May 2004 (vol. 165, no. 22), 341.

Heatherley, Ana Nez. *Healing Plants: A Medical Guide to Native North American Plants and Herbs*. New York: Lyons Press, 1998.

Henry, Brian. Personal communication, Aboriginal arrival in North America, Assembly of First Nations, Ottawa. 5/15/2009.

Herity, Michael, and George Eogan. *Ireland in Prehistory*. New York: Routledge, 1996.

Herrick, James W. *Iroquois Medical Botany*. Syracuse: Syracuse University Press, 1995.

Hillier, Harold. *The Hillier Manual of Trees and Shrubs*. Newton Abbot, Devon, England: David and Charles Redwood, 1992.

Hosie, R. C. *Native Trees of Canada*. Ottawa: Department of Fisheries and Forestry, 1969.

Howes, F. N. *Nuts: Their Production and Everyday Use*. London: Faber and Faber, 1948.

———. *Plants and Beekeeping*. London: Faber and Faber, 1979.

Jaynes, Richard A. *Nut Tree Culture in North America*. Hamden, Conn.: Northern Nut Growers' Association, 1979.

Kingsbury, John M. *Poisonous Plants of the United States and Canada*. Englewood Cliffs, N.J.: Prentice-Hall, Inc., 1964.

Klarreich, Erika. "Computation's New Leaf, Plants May Be Calculating Creatures." *Science News*, February 2004 (vol. 165, no. 8), 123–24.

Klots, Alexander B. *A Field Guide to Butterflies of North America, East of the Great Plains*. Boston: Houghton Mifflin, 1951.

Krieger, Louis C. *The Mushroom Handbook*. New York: Dover, 1967.

Krochmal, Arnold, and Connie Krochmal. *The Complete Illustrated Book of Dyes from Natural Sources*. New York: Doubleday, 1974.

Lee, Robert Edward. *Phycology. 2nd ed.* Cambridge: Cambridge University Press, 1995.

Lellinger, David B. *A Field Manual of Ferns and Fern Allies of the United States and Canada*. Washington, D.C.: Smithsonian Institution Press, 1985.

Lewis, Walter H., and P. F. Elvin-Lewis. *Medical Botany: Plants Affecting Man's Health.* Toronto: John Wiley and Sons, 1979.

Liberty Hyde Baily Hortorium. *Hortus Third: A Concise Dictionary of Plants Cultivated in the United States and Canada.* New York: Macmillan, 1976.

Little, Elbert L. *Trees.* New York: Alfred A. Knopf, 1980.

Megan, Ruth, and Vincent Megan. *Celtic Art.* London: Thames and Hudson, 1999.

Michl, Josef. "Mini Motors, Synthetic Molecule Yields Nanoscale Rotor." *Science News,* March 2004 (vol. 165, no. 12), 180.

Milius, Susan. "The Social Lives of Snakes from Loner to Attentive Parent." *Science News,* March 2004 (vol. 165, no. 13), 200–201.

———. "Thoroughly Modern Migrants: Moths and Butterflies—Round Trip Tickets Not Necessary." *Science News,* June 2004 (vol. 165, no. 26), 408–10.

———. "Travels with the War Goddess." *Science News,* May 2004 (vol. 165, no. 22), 344–46.

———. "Warm-Blooded Plants?" *Science News,* December 2003 (vol. 164, no. 24), 379–81.

Miller, Timothy L. "A Portrait of Pollution, Nation's Fresh Water Gets a Check-up." *Science News,* May 2004 165, 325–26.

Mullins, E. J., and T. S. McKnight. *Canadian Woods: Their Properties and Uses.* Toronto: University of Toronto Press, 1981.

Mumby, Peter J. "Mangrove Might, Nearby Trees Boost Reef-Fish Numbers." *Science News,* February 2004 (vol. 165, no. 6), 85–86.

Myers, Norman. *Gaia: An Atlas of Planet Management.* New York: Doubleday, 1984.

Peters, Annette. "Heavy Traffic May Trigger Heart Attacks." *Science News,* November 2004 (vol. 166, no. 20), 316.

Peterson, Roger Tory, and Margaret McKenny. *A Field Guide to Wild Flowers of Northeastern and North-central North America.* Boston: Houghton Mifflin, 1968.

Phillips, Roger, and Martyn Rix. *Perennials. 2 vols.* New York: Random House, 1991.

Pirone, P. P. *Tree Maintenance.* 6th ed. Oxford: Oxford University Press, 1988.

Pryer, Kathleen M. "A Frond Farewell, Genes Hint That Ferns Prolifer-
ated in Shade of Flowering Plants." *Science News,* April 2004 (vol. 165,
no. 14), 214.

Rackham, Oliver. *The Illustrated History of the Countryside.* London: George
Weidenfeld and Nicolson, 1994.

Raloff, Janet. "Danger on Deck." *Science News,* January 2004 (vol. 165, no. 5),
74–76.

———. "Dead Waters: Massive Oxygen-Starved Zones Are Developing Along
the World's Coasts." *Science News,* June 2004 (vol. 165, no. 23), 360–62.

Ramsayer, Kate. "Infrasonic Symphony: The Greatest Sounds Never Heard."
Science News, January 2004 (vol. 165, no. 2), 26–28.

Reeves, Randall. Personal communication on the great whales of Sakhalin
Island, May 2009.

Rupp, Rebecca. *Red Oaks and Black Birches: The Science and Lore of Trees.* Pow-
nal, Vt.: Storey Communications, 1995.

Schopmeyer, C. S. *Seeds of Woody Plants in the United States.* Washington, D.C.:
Forest Service, U.S. Department of Agriculture, 1974.

Small, Ernest, and Paul M. Caitling. *Canadian Medical Crops.* Ottawa: National
Research Council of Canada, 1999.

Smith, Russell J. *Tree Crops: A Permanent Agriculture.* New York: Devin-Adair,
1953.

Stuart, Malcolm. *The Encyclopedia of Herbs and Herbalism.* London: Orbix,
1979.

Taylor, Kathryn S., and Stephen F. Hamblin. *Handbook of Wildflower Cultiva-
tion.* New York: Macmillan, 1963.

Travis, John. "All the World's a Phage: Viruses That Eat Bacteria Abound—And
Surprise." *Science News,* July 2003 (vol. 164, no. 2), 26–28.

Uhari, Matti K. "A Sugar Averts Some Ear Infections." *Science News,* October
1998 (vol. 154, no. 18), 287.

Waldron, G. E. *The Tree Book: Tree Species and Restoration Guide for the
Windsor-Essex Region.* Windsor, Ont.: Project Green, 1997.

Wang, Ying Qiang. "A New Slimy Method of Self-Pollination." *Science News,*
September 2004 (vol. 166, no. 12), 190.

Wickens, G. E. *Edible Nuts.* Rome: Food and Agriculture Organization of the
United Nations, 1995.